Studies in Applied Philosophy, Epistemology and Rational Ethics

Volume 78

Mireille Hildebrandt, Erasmus MC, Rotterdam, The Netherlands

Michael H. G. Hoffmann, School of Public Policy, Georgia Institute of Technology, Atlanta, USA

Jeroen van den Hoven, Faculty of Technology, Policy and Management, Delft University of Technology, Delft, The Netherlands

Woosuk Park, Humanities and Social Sciences, KAIST, Daejeon, Korea (Republic of)

Gerhard Minnameier, Goethe-Universität Frankfurt am Main, Frankfurt, Germany

Yukio Ohsawa, School of Engineering, The University of Tokyo, Tokyo, Japan

Sami Paavola, Faculty of Educational Sciences, University of Helsinki, Helsinki, Finland

Alfredo Pereira, Institute of Biosciences, Universidade de São Paulo, São Paulo, Brazil

Luís Moniz Pereira, Faculdade de Ciências e Tecnologia, Universidade Nova de Lisboa, Caparica, Portugal

Ahti-Veikko Pietarinen, Department of Philosophy, University of Helsinki, Helsinki, Finland

Demetris Portides, Department of Classics and Philosophy, University of Cyprus, Nicosia, Cyprus

Dagmar Provijn, Centre for Logic and Philosophy, Ghent University, Ghent, Belgium

Joao Queiroz, Institute of Arts and Design, Universidade Federal de Juiz de Fora, Juiz de Fora, Brazil

Athanassios Raftopoulos, Psychology, University of Cyprus, Nicosia, Cyprus

Ferdie Rivera, Department of Mathematics and Statistics, San Jose State University, San Jose, USA

Colin T. Schmidt, Ensam ParisTech and Le Mans University, Laval, France

Gerhard Schurz, Department of Philosophy, DCLPS, Heinrich Heine University, Düsseldorf, Germany

Nora Schwartz, Department of Humanities, Universidad de Buenos Aires, Buenos Aires, Argentina

Cameron Shelley, Centre for Society, Technology and Values, University of Waterloo, Waterloo, Canada

Frederik Stjernfelt, Center for Semiotics, University of Aarhus, Aarhus C, Denmark

Mauricio Suárez, Logic and Philosophy of Science, Complutense University, Madrid, Spain

Peter-Paul Verbeek, Faculty of Science, University of Amsterdam, Amsterdam, The Netherlands

Riccardo Viale, Economics, Management and Statistics, University of Milano-Bicocca, Milan, Italy

Marion Vorms, Pantheon-Sorbonne University, Paris, France

Donna E. West, Department of Modern Languages, State University of New York College at Cortland, Cortland, NY, USA

Studies in Applied Philosophy, Epistemology and Rational Ethics (SAPERE) publishes new developments and advances in all the fields of philosophy, epistemology, and ethics, bringing them together with a cluster of scientific disciplines and technological outcomes: ranging from computer science to life sciences, from economics, law, and education to engineering, logic, and mathematics, from medicine to physics, human sciences, and politics. The series aims at covering all the challenging philosophical and ethical themes of contemporary society, making them appropriately applicable to contemporary theoretical and practical problems, impasses, controversies, and conflicts.

Book proposals can be sent to the Series Editor, Prof. Lorenzo Magnani at lmagnani@unipv.it. Please include: A short synopsis of the book or the introduction chapter A Table of Contents The CV of the lead author/editor(s) For more information, please contact the Editor-in-Chief at lmagnani@unipv.it

Our scientific and technological era has offered "new" topics to all areas of philosophy and ethics – for instance concerning scientific rationality, creativity, human and artificial intelligence, social and folk epistemology, ordinary reasoning, cognitive niches and cultural evolution, ecological crisis, ecologically situated rationality, consciousness, freedom and responsibility, human identity and uniqueness, cooperation, altruism, intersubjectivity and empathy, spirituality, violence. The impact of such topics has been mainly undermined by contemporary cultural settings, whereas they should increase the demand of interdisciplinary applied knowledge and fresh and original understanding. In turn, traditional philosophical and ethical themes have been profoundly affected and transformed as well: they should be further examined as embedded and applied within their scientific and technological environments so to update their received and often old-fashioned disciplinary treatment and appeal. Applying philosophy individuates therefore a new research commitment for the 21st century, focused on the main problems of recent methodological, logical, epistemological, and cognitive aspects of modeling activities employed both in intellectual and scientific discovery, and in technological innovation, including the computational tools intertwined with such practices, to understand them in a wide and integrated perspective. **Studies in Applied Philosophy, Epistemology and Rational Ethics** means to demonstrate the contemporary practical relevance of this novel philosophical approach and thus to provide a home for monographs, lecture notes, selected contributions from specialized conferences and workshops as well as selected Ph.D. theses. The series welcomes contributions from philosophers as well as from scientists, engineers, and intellectuals interested in showing how applying philosophy can increase knowledge about our current world. Indexed by SCOPUS, zbMATH, SCImago, DBLP. All books published in the series are submitted for consideration in Web of Science.

Yaoli Du

Machines in Minds

Expressing Digital Rationality within the Space of Reasons

Springer

Yaoli Du
Institut für Philosophie
Technische Universität Braunschweig
Braunschweig, Germany

ISSN 2192-6255 ISSN 2192-6263 (electronic)
Studies in Applied Philosophy, Epistemology and Rational Ethics
ISBN 978-3-032-15841-3 ISBN 978-3-032-15842-0 (eBook)
https://doi.org/10.1007/978-3-032-15842-0

This Springer imprint is published by the registered company Springer Nature Switzerland AG
The registered company address is: Gewerbestrasse 11, 6330 Cham, Switzerland

Preface

This book is based on my doctoral dissertation which was accepted at the University of Leipzig in 2023. After my defence in 2024, the manuscript has been revised and supplemented for publication.

The research theme of this book is a philosophical inquiry into the possibilities of artificial intelligence to serve as rational agents in digital social practices and its conditions. This inquiry re-examines the relationship between minds and machines. On this basis, it explores the intersubjective origins and normative guidance of what is conceived as a thinking machine. At a time of extremely rapid advances in digitalisation and AI technologies, this work is urgent. Although the original intent of this research was to devote a debate within philosophy on what truth is and how it operates as a kind of commitment to guide us in our social practices as rational beings. This inquiry inevitably involves the core areas of philosophy such as epistemology, ontology, philosophy of mind and language, as well as more recent research perspectives like philosophy of information and philosophy of AI. Such a research project across perspectives and disciplines is profoundly hard. In proposing, launching, and completing it, I have received much generous help and expertise from a number of people. I would like to express my heartfelt thanks to those who have supported and helped me to bring this thesis to fruition.

The initial idea for this work emerged during my graduate studies at the University of Leipzig. On the one hand, the wide range of study and research programmes provided by the Institute of Philosophy allowed me to gain insight into the critical reception of German idealism by neopragmatism and inferentialism. In particular, the lectures and seminars offered by Pirmin Stekeler-Weithofer were of great help to me in understanding the connection between Hegel and debates in contemporary philosophy of mind and language. After completing my master's thesis under his supervision, I prepared my proposal for the doctoral project and submitted it to Kristina Musholt following his suggestions. I am deeply grateful for his kind advice. The lecture series held by Robert Brandom during his 2015–2019 visits to Leipzig gave me the invaluable opportunity to directly benefit from his most recent findings and thinkings. On the other hand, through the programme offered by Hans-Gert Gräbe at the Institute for Informatics, I gained access to semantic technologies and their unique role in digital practices, while developing a deeper understanding of the philosophical issues underlying data and information science. I appreciate the interdisciplinary exchange opportunities to widen my perspectives and for his valuable feedback on my work after I completed the first draft of my doctoral thesis.

My deepest appreciation goes to my supervisor, Kristina Musholt, for her tremendous help and patience throughout the entire project, from my initial proposal to the finalisation of the thesis. She gave me numerous valuable comments and pertinent criticisms on my work. In particular, from the perspectives of cognitive anthropology and philosophy of psychology, she encouraged me to explore in depth the comparability of the conditions of human intelligence and the conditions of machine intelligence. I have also benefited a lot from her helpful advice and guidance on how to keep philosophical discernment and interdisciplinary perspectives. These always remind me of how philosophy can engage with the current practical issues in its uniquely theoretical yet discursive way. I am also incredibly grateful to Wolfgang Neuser for kindly accepting my request to be the second examiner for my dissertation. His work inspired me on the possibility and necessity to understand the digital society from a knowledge society standpoint. He generously motivated me to explore further how normative statuses could be assigned to subjects from their intersubjective origins. With their help, this thesis finds theoretical and practical anchors. The findings of my dissertation suggest that, in very general terms, AI agents' potential to act in a rational and responsible manner depends on human agents' rational and responsible actions in digital social practices. Artificial rational agents are holders of normative statuses who can express themselves in shared digital practices. It ultimately advocates diverse and responsible digital practices, in a way that general human users and developers can implement in order to allow for reasonable and responsible human-AI interactions.

During my doctoral studies, I had the opportunity to present my work regularly at the research colloquium Cognitive Anthropology. Many thanks to Nikolaos Psarros, Sascha Benjamin Fink, Nguyen Trung Hau, Robert Reimer, to name just a few, for their constructive feedback and insightful discussions. I particularly appreciate Katrin Felgenhauer for her helpful remarks and writing suggestions. And special thanks to Nadine Schumann for her profound discussions and sharing of experiences with me. Together we have co-authored and published some papers on related topics, which have motivated me to stay committed to my research theme.

Moreover, I would like to thank the SFB 991 research project at the Heinrich Heine University Düsseldorf for providing me with a visiting fellowship in 2019–2020, with special appreciation to Gottfried Vosgerau and Sebastian Löbner for their helpful advice. Further, I would like to thank Thomas Kater and Center for Teacher-Training and School Research (ZLS) at the University of Leipzig for offering me a teaching assistantship in the final stage of writing my dissertation, which allowed me to complete my dissertation while sharing my research and experience in graduate seminars.

Since 2024 I have been working as a research associate at the Technical University of Braunschweig, where I reported my thesis on the research colloquium and received plenty of very helpful feedback. I hereby thank Nicole Karafyllis, Tobias

Endres, Domenico Schneider, Hans-Christoph Schmidt am Busch, Claudia Wirsing, and Nina Reinfelder. Particular appreciation to Nicole Karafyllis for her kind help and precious advice, which allowed me to engage in diverse academic exchanges while working on my research and teaching, as well as completing my defence and revising my thesis for publication. Special thanks to Magnani Lorenzo, Editor-in-Chief of the book *Studies in Applied Philosophy, Epistemology and Rational Ethics*, and the anonymous reviewers for their valuable feedback on the manuscript, as well as to Technical Editor Leontina Di Cecco for her kind support in bringing it to publication.

I am indebted most to my family, especially my mother, Xiaowen Du, for her untold support, trust, and love. I would also like to thank my partner Ken Pierre Kleemann, who has been encouraging me since the project was still a fledgling idea and has been always there for me.

This book is dedicated to all those who have helped me make it possible, and to you who will read it.

Leipzig, Germany Yaoli Du

Competing Interests The author has no competing interests to declare that are relevant to the content of this manuscript.

If everything is idea, then there is no owner of ideas. [...] If there is no owner of ideas then there are also no ideas, for ideas need an owner and without one they cannot exist.[1]

Frege, Thought (1918)

As compared with the other animals man lives not merely in a broader reality; he lives, so to speak, in a new dimension of reality.[2]

Cassirer, An Essay on Man: An Introduction to a Philosophy of Human Culture (1944)

[1] Gottlob Frege, "Thought [1918]," in *The Frege Reader*, ed. Michael Beaney (Oxford: Blackwell, 1997), 339.

[2] Ernst Cassirer, *An Essay on Man: An Introduction to a Philosophy of Human Culture* (New Haven: Yale University Press, 1944), 42f.

Contents

1 Introduction 1
1.1 The Rise of Computational Functionalism and Its Limitations 1
1.2 A Normative Turn from Functional States to Intersubjective Statuses 11
Bibliography 15

2 How Minds Relate to Machines 19
2.1 Category Mistakes 19
2.1.1 The Mind-Body Problem in Spatial Metaphor 20
2.1.2 The Category Mistake of Embodying Mind 22
2.1.3 The Category Mistake of Encoding Mind 25
2.2 Computer Analogies and Challenges 27
2.2.1 Cybernetic Machinery vs. Artificial Intelligence 28
2.2.2 Symbolic AI, the Computational Theory of Mind, and the Semantic Issue 32
2.2.3 Sub-Symbolic AI, Connectionism, and the Unexplainable Issue 38
2.3 Rethinking the Relation Between Minds and Machines in Logical Space 45
Bibliography 51

3 A Novel Space for Reframing Mind-Machine Relations 55
3.1 Data, Information, Knowledge 56
3.1.1 Data and Facts 56
3.1.2 Information and Reality 58
3.1.3 Knowledge and Authorities 61
3.2 Semantic (Web) Technologies in Digital Social Practices 67
3.2.1 Semantic Web Architecture and Tools 69
3.2.2 Knowledge Engineering for Structuring Data and Metadata 75

3.3 Linked Cyberspace as a Materialized Logical Space 83
3.3.1 The Order of "Things" in Cyberspace and "Material Intentionality" 83
3.3.2 A Materialized Logical Space for Information Processing Agents 87
Bibliography 92

4 **Triandic Interaction, Inter-Subjectivity, Inter-Objectivity** 97
4.1 From the Imitation Game to a Play of Interaction 97
4.2 Recursive Patterns of Meaning-Truth-Beliefs for Interpretation 100
4.2.1 Interpreting Meaning with Truth-Conditional Semantics 101
4.2.2 Indeterminate Translation and Radical Interpretation 104
4.2.3 Principle of Charity and Commitment to an Inter-Subjective Reality 109
4.3 Patterns of Meaning-Making in Triadic Interaction 114
4.3.1 Inter-Objective Foundation of Knowledge 114
4.3.2 The Formation of Pragmatic Action Patterns Through Triangulation 120
Bibliography 125

5 **Rational Agency in the Normative Space of Reasons** 127
5.1 In-Formation of the Normative Space of Reasons 127
5.1.1 Encoding Material Inference in Language Use 128
5.1.2 Decoding Rationalist Commitments to Modality and Normativity 132
5.2 Operative Intentionality in Rational Agency 137
5.3 Inferential Articulation of Norm-Making 141
5.3.1 Recognition of Normative Statuses and Reciprocal Authorization 142
5.3.2 Expressing Practical Reasoning with "Pragmatic Metavocabularies" 150
Bibliography 158

6 **Expressiving Rational Agency with Language Models** 161
6.1 General-Purpose Representations Trained on Web Resources 161
6.2 Human-Like Representations and Human-Like Biases 166
6.3 Alignment in Pluralistic Triadic Interactions 169
6.4 Expressive Rational Agency in Digital Discursive Practices 173
Bibliography 178

7 **Forming Rational Agents in Digital Practices** 181
7.1 Structuring Shared Reality in Linked Cyberspace of Reasons 181
7.2 Forming Pragmatic Information to Afford Interpretable Interactions 184
7.3 Authorizing Artificial Agents to Practice Expressing Norm-Making 185

Chapter 1
Introduction

Can machines think? With the widespread digitalization and application of artificial intelligence (AI) technologies today, the question originally posed by Turing worth revisiting. This book is devoted to examining whether and under what conditions conceptions of AI can be rational. It will shed light on a normative shift in understanding how machines and minds stand in relation to each other. This shift will be argued as necessary for the developmental goals of AI. Before delving into the necessity of the shift, this chapter introduces the topic by elucidating how AI has become both a research object and methodology in cognitive science, and how its conceptual foundations, underlying hypotheses, and challenges are intrinsically related to philosophy, especially mathematical logic and philosophy of mind. It will also outline the structure of the book to make it easier for readers to navigate.

1.1 The Rise of Computational Functionalism and Its Limitations

AI, or what implements it, computing machines, as a research discipline and development goal, is inextricably linked to the central hypothesis of functionalism, namely that it is possible to imitate thinking processes with machines, even if minds are not identical to machines. The explicit idea that it is possible to automatically derive and prove arithmetic formulas with machines, and even generate universal knowledge, can be traced back at least to Leibniz. Leibniz holds that all knowledge is compositional like mathematical knowledge and looks for laws of combinations.[1]

[1] For example, "1 + 3 = 4" can be understood as the combination of the given numbers 1 and 3 through rules of calculation to generate a new arithmetic concept in terms of mathematical knowledge. In this way, complex arithmetic concepts are formed from the combination of simple arithmetic concepts. Gottfried W. Leibniz, "Dissertation on the Art of Combinations 1666 (Selections),"

Y. Du, *Machines in Minds*, Studies in Applied Philosophy, Epistemology and Rational Ethics 78, https://doi.org/10.1007/978-3-032-15842-0_1

His crucial idea is that if the method of deriving mathematical knowledge can be extended to other domains of knowledge, then all human knowledge can be derived and combined with the help of a universal symbolic system.[2] Accordingly, knowledge from different domains can be represented and coherently linked to each other in a universal symbolic system. Leibniz thus envisions a universal symbolic language "*characteristica universalis*" and a universal calculation for combining characters "*calculus ratiocinator*", which would be able to perform the rules of reasoning of a symbolic language in a general way.[3] Based on this mathematically inspired assumption, a general representation of knowledge acquired by mechanical operations according to calculating laws is thought to be possible.[4]

It is an attractive assumption for philosophers and mathematicians that there are universal laws of reasoning and acquiring knowledge. Boole regards the laws of algebra as "laws of thought" that underlie reasoning, but argues that they derive from mental faculties.[5] In the corresponding "psychologism dispute" of the nineteenth century, the question of what grounds logical reasoning is at the centre of the debate.[6] Psychologism is the position that views reasoning as an intrinsic mental process and holds that logical rules derive from mental laws, hence the study of logical reasoning is considered a branch of psychology.[7] In this position mathematical laws typically serve as a mediator for mental faculties to derive logical reasoning. Yet if mathematical laws need to be examined by empirical science as

in *Philosophical Papers and Letters*: A Selection, ed. Leroy E. Loemker (Dordrecht: Kluwer Academic Publishers, 1989), 73–84.

[2] Gottfried W. Leibniz, "On the Method of Universality [1674]," in *Leibniz: Selections*, ed. Philip P. Wiener (New York: Charles Scribner's Sons, 1951), 4.

[3] Gottfried W. Leibniz, "Letter to Nicolas Remond, 1714-15," in *Philosophical Papers and Letters*, 654.

[4] Gottfried W. Leibniz, "Preface to the General Science [1677]," in *Leibniz: Selections*, 15.

[5] Boole states that the purpose of his research is to provide a reliable mathematical basis for the activity of thought in performing logical reasoning, and to express these thoughts in a symbolic language. But Boole rejects metaphysical explanations for the ability to perform arithmetic operations and logical reasoning and emphasizes that the nature and practical application of these laws should be studied using the methods of empirical science. d. George Boole, *An Investigation of The Laws of Thought, on Which are Founded Mathematical Theories of Logic and Probabilities* (New York: Dover, 1854), 29ff.

[6] At the end of the nineteenth century, the birth of new disciplines such as experimental psychology led to a philosophical movement that rejected the so-called dogmatic metaphysics. Epistemological issues were thought to be solvable through empirical science.

[7] Psychologism is represented by John S. Mill, Theodor Lipps, and Benno Erdmann, to name a few. According to Brockhaus' distinction between weak and strong psychologism, the weak version holds that "logical principles are inherent in the laws of mind, and uncoverable by psychological methods", while the strong version holds that "logic is literally a branch of psychology". Richard R. Brockhaus, "Realism and Psychologism in 19th Century Logic," *Philosophy and Phenomenological Research* 51, no. 3 (1991): 495.

psychological laws, then the legitimacy of mathematics as a general tool is called into question.[8]

In exploring the foundations of arithmetic, Frege questions the legitimacy of psychological laws as foundations in general. His insight is that numbers and other mathematical propositions should be understood as concepts bound by logical norms.[9] Mathematical operations are no longer based on individual mental processes, but on publicly usable concepts. Frege argues that Boole understood mathematical notation as a formal system of writing and reasoning but ignored how material reality could be conceptualized by symbols and their relations.[10] A strict distinction should be made between concepts and the objects in material reality that are conceptualized. In his anti-psychologism account, Frege points out that the problem with psychologism is that it confuses the psychological realm with the realm of objectivity, since the mental act of reasoning and the object being reasoned about arc usually mixed up here.[11] He argues that the objective realm is a third realm in addition to the psychological and physical realms, to which the normative concepts and logical rules belong.[12] The objective realm can be understood as a logical and normative space in which the foundations of arithmetic lie, for "number is extension of concept" and "arithmetic is a development of logic".[13] Accordingly, the study of operations and reasoning is no longer an inquiry into the "laws of thought" by psychological methods, but an exploration of the "laws of truth" in the logical and normative space beyond the psychological and physical realms.[14]

Investigation of "laws of truth" triggered not only the well-known mathematical crisis but also opened the way for the study of symbolic systems and modern

---

[8] Opponents of Psychologism argue that the psychologistic view of reasoning as mental laws can lead to relativism and skepticism. For if logical reasoning is essentially a psychological operation derived from mental laws, then the basis of logical reasoning can vary according to different individual experiences and cultural differences. As Brockhaus points out, "[i]ronically, although psychologism styles itself as providing logic with an objective foundation, it introduces precisely the anthropomorphic relativism which Frege, Russell, Husserl and Wittgenstein were to hold in contempt". Ibid., 498.

[9] Gottlob Frege, *The Foundations of Arithmetic, A logico-mathematical enquiry into the concept of number*, trans. J.L. Austin (New York: Harper Torchbooks, 1960), xvi.

[10] Gottlob Frege, "Boole's Logical Culculus and the Begriffsschrift. [1880/81]," in *Posthumous Writings*, eds. Hans Hermes, Friedrich Kambartel, and Friedrich Kaulbach (Chicago: University of Chicago Press, 1979), 13.

[11] Frege, *The Foundations of Arithmetic*, 38.

[12] Ibid., 35.

[13] Ibid., 99.

[14] In his late essay Thought (1918), Frege states that the task of logic is "to discern the laws of truth". He notes that the word "laws" can be used to express, on the one hand, man-made regulations such as laws and orders, and on the other hand, to describe the "laws of nature". The mistake of those who hold on to psychologism is in drawing a direct analogy between the "laws of thought" and the "laws of nature". Instead, Frege holds that logical reasoning takes place in the objective realm that links the psychological and actual realms. The mathematical symbols and true propositions in this realm are the thoughts as well as tools that people share. Gottlob Frege, "Thought [1918]," in *The Frege Reader*, ed. Michael Beaney (Oxford: Blackwell, 1997), 325ff.

computer science. The crisis began with a series of paradoxes arising from attempts to systematically establish the foundations of mathematics, including approaches from set theory and mathematical logic. Russell uses the term "self-referential" problem to group together this series of paradoxes, and sums up the problem as failing to hierarchize the logical structure of language.[15] At the heart of the issue is how to self-consistently express the "law of truth" in symbolic language. To prevent paradoxes, mathematicians attempt to axiomatize mathematics and to lay a consistent foundation for mathematics.[16] Yet, Gödel's incompleteness theorems reveal the impossibility of complete and consistent axiomatization, because there are undecidable sentences in a consistent formal system that can neither be proven nor disproven.[17] Godel's proof of incompleteness theorems inspires several relevant studies of the limitations of formal symbolic systems. On the one hand, Tarski proposes the undefinability theorem in terms of the semantics of formal languages, which states that a forma language system cannot define semantic truth for the concepts within it.[18] This prompts Tarski to redefine the concept of truth as a formal relation by distinguishing between meta- and object languages.[19] On the other hand, the subtlety of Gödel's proof is the use of numbers as tools to guide mathematical derivations, so that logical reasoning can also be expressed numerically. This numerical encoding method is further adopted by Church and Turing to deal with issues of "effective calculability" or "effective computability" in response to the "Entscheidungsproblem".[20]

---

[15] Bertrand Russell, "Mathematical Logic as Based on the Theory of Types," *American Journal of Mathematics* 30, no. 3 (1908): 224.

[16] David Hilbert, "Mathematische Probleme," in *Göttinger Nachrichten* (1900):253–297, also in *Archiv der Mathematik und Physik* 3, no.1 (1901): 44–63 and 213–237. English trans. Mary Winton Newson, In *Bulletin of the American Mathematical* Society 8 (1902): 437–479. Hilbert proposes 23 unsolved problems on the foundations of mathematics.

[17] Kurt Gödel, "Über formal unentscheidbare Sätze der Principia Mathematica und verwandter Systeme I," *Monatshefte für Mathematik* 38, no.1 (1931):173–198.

[18] Alfred Tarski, "The Concept of Truth in Formalized Languages," in *Logic, Semantics, Metamathematics*, ed. Alfred Tarski (Oxford: Oxford University Press, 1936), 187.

[19] Ibid. Tarski's theory of truth for formal language was later extended by Davidson to natural language and the study of formal semantics. I will return to and elaborate on this point in Chap. 4.

[20] Briefly, the Entscheidungsproblem asks whether there is a universal procedure that can determine the validity of any given logical expression in first-order logic in a finite number of steps. It is a query about the derivability of first-order logic. This problem was posed in Hilbert's unsolved problems and was attempted to be solved by Gödel, Church and Turing around the same period. Kurt Gödel, "On Undecidable Propositions of Formal Mathematical Systems (1934)." Reprinted in *Collected Works. I: Publications 1929–1936*, eds. Solomon Feferman et al. (Oxford: Oxford University Press, 1986), 346. Alonzo Church, "An Unsolvable Problem of Elementary Number Theory," *American Journal of Mathematics* 58, no. 2 (1936): 356. Alan Turing, "On Computable Numbers, with an Application to the Entscheidungsproblem," *Proceedings of the London Mathematical Society* s2–42, Issue 1, (1937): 231. The exploration of this problem was transferred by Turing from mathematics to programming and continues into today's work in information technology and software engineering. A more detailed discussion about the numeric encoding can be seen in Klaus Mainzer, *Die Berechnung der Welt. Von der Weltformel zu Big Data* (München: C.H. Beck, 2014), 89 ff.

Turing defines the term computability by using an analogy between a person who is computing a number and an abstract machine model that can mimic this activity.[21] Unlike its use today, the term "computer" was used at the time to describe the employed people who perform computing.[22] Turing notes that "[t]he behavior of the computer at any moment is determined by the symbols which he is observing, and his 'state of mind' at that moment", and thus argues that the process of calculation by a human computer can be decomposed into state changes in processing mathematical symbols.[23] Namely, in calculating, read-write operations and changes in mental states that follow given rules can be regarded as changes in states of symbol configuration, which can be mechanically simulated. Calculations or symbolic processing are no longer understood as following "laws of thought" that are rooted in mental faculties, but as procedural operations. The step-by-step procedure of how a human computer operates symbols to perform calculations can thus be modelled by an abstract logical computing machine, the Turing machine. This is Turing's main thesis that any number, or more generally any sequence of symbols, which is computable for a human is also computable by a Turing machine.[24]

The concept of the Turing machine allows the equivalent conversion of finding a universal procedure in the "Entscheidungsproblem" to a search for a finite mechanical procedure that can determine a given formula in first-order logic.[25] A universal procedure is accordingly defined as a universal Turing machine that can execute the computing process of any Turing machine, that is, operate any Turing machine as a program.[26] Programs can therefore be understood as structured or organized physical patterns that can be read and executed by a machine.[27] The machine that can read and run programs, in turn, requires a more general structure. The distinction between programs and machines is not absolute.[28] A Turing machine can be both the program that carries these instructions and the machine that executes them. Turing eventually proves that an axiomatic first-order predicate calculus system can derive

[21] Turing, "On Computable Numbers, with an Application to the Entscheidungsproblem," 230f.

[22] For a history of human computers, especially the contributions by women who made up the majority, see Claire L. Evans, *Broad band: the untold story of the women who made the Internet* (New York: Penguin Putnam, 2018).

[23] Turing, "On Computable Numbers, with an Application to the Entscheidungsproblem," 250.

[24] For a more systematic explanation, see Jack Copeland, *The Essential Turing: Seminal Writings in Computing, Logic, Philosophy, Artificial Intelligence, and Artificial Life: Plus the Secrets of Enigma* (Oxford: Clarendon, 2004), 41ff.

[25] The "Entscheidungsproblem" problem can now be converted into the printing problem, i.e., whether there is a universal machine that can determine an arbitrary machine to print a given symbol. The printing problem can also be interpreted as the halting problem, which asks a general algorithm for determining whether an arbitrary program and its inputs can end its run without falling into a loop. The term "halting problem" is firstly introduced by Davis. Martin Davis, *Computability and Unsolvability* (New York: McGraw-Hill, 1958), 70.

[26] Turing, "On Computable Numbers, with an Application to the Entscheidungsproblem," 241f.

[27] Peter Suber, "What is Software?" *The Journal of Speculative Philosophy* 2, no. 2 (1988): 104ff.

[28] Suber argues hardware is materialized instructions for computation, so that it is also essentially program or software. Ibid., 90.

formulas using the mechanical procedure he defines, but there is no universal machine that can determine in finite steps whether a given formula is provable.[29] In this particular way, he reveals the possibilities and limitations of performing calculating operations through symbolic manipulation on machine models.[30] Based on the notion of the universal machine, Turing starts to consider the relationship between the potential of universal machines and the concept of intelligence in general, and to conceive of "possible ways in which machinery might be made to show intelligent behavior".[31]

In Turing's 1948 report entitled *Intelligent Machinery*, he further draws analogies between human and machine intelligence, not only between human neural organization and machine organization, but also between the education of humans and the training of machines.[32] These considerations profoundly influenced later AI research. On the one hand, it looks forward to the possibility of simulating human intelligence with machine models. On the other hand, it prospects the impact of human attitudes towards intelligence when interacting with machines. As Turing suggests at the end of his 1948 report, intelligence can be understood "as an emotional concept" that depends on the observation and attribution of the behavior of others.[33] He conceives a test between two human chess players and a paper-machine that plays chess through an operator.[34] If all three "players" play chess at the same time, one of the human players has difficulty determining who is her opponent. This thought experiment is a prelude to his conception of the more general imitation game in his 1950 paper, in which he refines this thought experiment into the famous Turing test to replace and strategically answer the question "Can machines think?".[35]

His 1950 paper *Computing Machinery and Intelligence*, published in the philosophical journal *Mind*, is more a reflection on the potential of machines and the

[29] Turing, "On Computable Numbers, with an Application to the Entscheidungsproblem," 252.

[30] Copeland sees two aspects of Turing's contributions to the development of the modern computer, namely, "[t]he idea of controlling the function of a computing machine by storing a program of symbolically encoded instructions in the machine's memory" and "[h]is demonstration [...] that, by this means, a single machine of fixed structure is able to carry out every computation that can be carried out by any Turing machine whatsoever, i.e. is universal." Copeland, *The Essential Turing*, 15.

[31] Alan Turing, "Intelligent Machinery," 3.

[32] Ibid., 14ff.

[33] Ibid., 23.

[34] Ibid.

[35] The refined imitation game is described as follows: "The new form of the problem can be described in terms of a game which we call the "imitation game." It is played with three people, a man (A), a woman (B), and an interrogator (C) who may be of either sex. The interrogator stays in a room apart from the other two. The object of the game for the interrogator is to determine which of the other two is the man and which is the woman [...].

We now ask the question, "What will happen when a machine takes the part of A in this game?" Will the interrogator decide wrongly as often when the game is played like this as he does when the game is played between a man and a woman? These questions replace our original, "Can machines think?" Alan Turing, "Computing Machinery and Intelligence," *Mind,* New Series 59, no. 236 (1950): 433f.

concept of intelligence from the perspective of theoretical computer science, than a strict computer engineering article. Turing argues that what might be considered a "thinking machine" would be a "digital computer" that could analogically simulate the decomposed human computing process with three basic components: (1) store, (2) executive unit, and (3) control.[36] A store is used to record information. It is equivalent to papers used by a human computer, on which one can write down calculation results or print the "book of rules". The executive unit performs arithmetic operations according to the rules. It is like a pen that moves and marks according to attention. And the control ensures that these rules are followed in the correct order in the "book of rules", which can correspond both to the rules in a human computer's memory and to the instruction tables that are applied to guide operations.[37] Correspondly, constructing complex operations into instruction tables is programming.[38] Such a digital computer could implement a universal Turing machine, as long as it has fast enough speed and large enough storage capacity to run a vast enough number of programs. In this way, the question "Can machine think?" can be transformed into the question "Are there imaginable digital computers which would do well in the imitation game?".[39] The "imitation game" is described as follows:

> The new form of the problem can be described in terms of a game which we call the "imitation game." It is played with three people, a man (A), a woman (B), and an interrogator (C) who may be of either sex. The interrogator stays in a room apart from the other two. The object of the game for the interrogator is to determine which of the other two is the man and which is the woman. [...] We now ask the question, "What will happen when a machine takes the part of A in this game?" Will the interrogator decide wrongly as often when the game is played like this as he does when the game is played between a man and a woman? These questions replace our original, "Can machines think?"[40]

The imitation game is cleverly designed so that the interrogator is not trying to identify which interlocutor is human (that makes no sense among three human players), but rather the gender of the interlocutor. In other words, Turing assigns specific roles to the players. Typically, social roles speak and act following some specific patterns, whether stereotypes or the need for responsibility. This ability to be "in character" is treated by Turing as a recognizable manifestation of intelligent machinery, similar to the discipline he mentioned in the 1948 report. Hence, the computer

[36] Ibid., 437. Turing writes that such pictorial analogy is not sufficient as a definition and may even lead to "danger of circularity of argument". Therefore, this analogy is only a simplified version of his argument for computable numbers.

[37] Ibid.

[38] Ibid., 438.

[39] This question has a more general form, namely, "Are there discrete state machines which would do well?" Turing emphasizes on the advantages of digital computers as "discrete state machines", because he argues that discrete state machines model changes in discrete states and can therefore simulate what is in fact continuous motion. States provide a possibility for quantifying and functionally simulating continuous qualities, which is the starting point of functionalism to simulate mental states with functional states. Ibid., 439, 442.

[40] Ibid., 433f.

in the imitation game is not imitating a thinking person, but is imitating someone who is imitating someone else.

As the question of the thinking machine changed from a metaphysical to an engineering and design problem, Turing lists nine typical objections and refuts each one. Among them, the fourth objection, "The argument from consciousness", and the sixth, "Lady Lovelace's Objection", are still the main arguments used today to deny the idea that computers can be as intelligent as humans. The consciousness argument holds that a machine can only be considered equivalent to a conscious brain if it can really "know" or "experience" emotional fluctuations.[41] In this regard, Turing admits that an extreme standpoint based on subjective conscious experience cannot be verified in the imitation game, but it may lead to solipsism.[42] A less extreme argument for consciousness, in Turing's view, could be tested entirely in the form of a dialogue in the imitation game. That is, the imitation game allows subjective experience to be interpreted by others. In a time when computers were far less powerful than today's AI technologies such as large language models, Turing postulated an artistic-looking dialogue between the interrogator and the machine, arguing that the difference between "merely artificially signal" and an artistic text would be only one of degree.[43]

In response to Lady Lovelace's argument that machines can only do what people know how to order it to do without its own creativity, Turing proposes the concept of "learning machines" and again compares machine learning to human education.[44] Learning machines are not trained to follow specific rules, but to generalize rules. He points out that the learning process of the machine may not be understood by the developers, which refutes Lady Lovelace's argument.[45] Turing's argument is well illustrated by today's widespread use of generative AI and its challenges. In addition, Turing makes another interesting argument about thinking, originality and making mistakes.[46] The learning of machines is not unconditional but based on training experience, and they are also likely to make mistakes. Yet the fact that either humans or machines follow rules taught by authorities or make mistakes does not demonstrate that they are unintelligent. Turing's approach sets the goal of computers that mimic human intelligence, including weak AI that mimics human intelligence as well the possibility of strong AI, i.e., fully achieving human intelligence, provided that human intelligence itself is based on imitation.[47]

As a research discipline originating in computer science, AI research has developed alongside interdisciplinary paradigms such as cybernetics and cognitive science. A series of conferences such as the Macy Conference and the Dartmouth

[41] Ibid., 445.

[42] Ibid., 446f.

[43] Ibid., 447.

[44] Ibid., 450, 459.

[45] Ibid., 459.

[46] Ibid.

[47] The discussion on strong and weak AI will be expanded in detail in Sect. 2.2.3.

Conference laid the tone for AI research.[48] How to simulate aspects of human intelligence has been approached from different theoretical entry points. On the one hand, the early stages of AI research employed computer technologies to simulate human intelligence for performing different types of tasks.[49] On the other hand, it became a common pursuit of interdisciplinary studies on cognition to provide a framework of principles and explanations for the intelligent operation of internal states and processes. Among the different explanatory frameworks of human cognition, functionalism upholds the analogy between human brains and machines. In distinction to the behaviorist treatment of internal processes as black boxes, or the reduction of mental states to human brain states in the identity theory, functionalism accounts for mental states as functional states.[50] The central idea is that mental states and processes can be understood as a system that plays characteristic causal roles and can be modelled by machines with state changes like the Turing machine. Accordingly, the same mental state can be actualised by multiple physical systems.[51] Turing's thesis of computability is reformulated here as treating both logical computer machine models and human brains as symbolic manipulation systems.

The analogous treatment of human brains and computing machines leads directly to the dominant functionalist account of mind in cognitive science, the representational computational theory of mind.[52] It focuses on how symbolic operating systems that play functional roles work and how they are formed.[53] According to Thagard's summary, the core assumption of this theory is that "mind (mental representation + computational procedures = thinking)" is analogous to "program (data structures + algorithms = running programs)".[54] The ability to represent external

[48] See Sect. 2.2.1 for a detailed discussion about the implications of these conferences.

[49] Cf. Antonio Lieto, *Cognitive Design for Artificial Mind* (Milton Park: Taylor & Francis, 2021).

[50] Hilary Putnam, "Psychological Predicates," in *Art, Mind and Religion*. eds. W. H. Captain and D. D. Merrill, 37–48 (Pittsburgh: University of Pittsburgh Press, 1967). Reprinted as "The Nature of Mental States," in *Mind, Language and Reality, Philosophical Papers*, Vol. 2, 429–440 (Cambridge: Cambridge University Press, 1975), 435ff.

[51] Ibid. 433f. In introducing functionalism, Putnam proposes a general machine model to model minds and calls it "Probabilistic Automaton", and he considers Turing machines to be a special kind of probabilistic automaton.

[52] Piccinini notes that functionalism and computationalism are not necessarily overlapping theories of the mind, because functionalism does not have to assume that the functional relation of internal states between input and output is computational. Due to a series of works by Putnam and Fodor, the controversy over functionalism became a controversy against computational functionalism. Cf. Gualtiero Piccinini, "Functionalism, computationalism, and mental states," *Studies in History and Philosophy of Science Part A* 35, no. 4 (2004): 811–33.

[53] Fodor offers a representationalist version of the computational theory of mind, for the machine-state functionalism cannot explain the productivity in natural language use and how the mind can systematically make sense of meaning. He argues that the cognitive process should be a computational process of composing mental representations into comprehensible causal sequences by using the syntax of the "language of thought". Jerry A. Fodor, *The Language of Thought* (Cambridge MA: Harvard University Press, 1975), 36. See Sect. 2.2.2 for further discussion of this hypothesis and its problems.

[54] Paul Thagard, *Mind: Introduction to Cognitive Science* (Cambridge, MA: MIT Press, 2005), 11.

reality with internal symbols, regarded as the main feature of thinking, is analogized to the ability to represent knowledge with structured databases. The corresponding idea for AI development is to model the rich semantic relations in mental representations with data structures and make them accessible to algorithms, i.e., well-defined syntactic rules of information processing that complete computations in a finite number of steps. Algorithms can be either fixedly programmed or acquired through data training. The symbolic AI approach and the sub-symbolic AI approach are debated in this context on how to obtain suitable algorithms.

The question of how algorithms are acquired relates to a more fundamental question of how mental representations and their information processing procedures are formed. Situated cognition as new paradigm has received increasing attention in this debate, which examines how material and environmental conditions affect the emergence and development of cognitive processes in terms of embodied, embedded, extended and enactive cognition.[55] With the broadening of theoretical frameworks, cognitive states and processes are no longer restricted within the brain, but are treated as distributed across bodily and environmental interactions such as the sensorimotor system, and even in the use of tools. But how such locally initiated cognitive processes give rise to a unified representation or high-level cognition becomes a new explanatory challenge.

As will be unfolded in Chap. 2, computational functionalist analogies and searches for mental attributes in internal states often fall into the trap of seeking the ghost in the machine. A rationalist and normative turn can transform modeling internal mental states and cognitive processes into modeling the interaction and communication between agents, as well as with their shared environment. This turn can provide a normative conceptual framework to complement the computational-representational theory of mind in order to help explain the irreducible features of mental states and representations and their social origins. Specifically, this turn can be captured from two aspects: understanding semantics in terms of pragmatics and understanding representation in terms of normative inference. These two aspects extend the notion of symbolic manipulation and mental representation from within individuals to between them and their shared world, breaking down the spatial metaphorical boundaries between the internal and the external with the enactment of practical norms.

---

[55] Cf. John Haugeland, "Mind embodied and embedded," in *Having thought: essays in the metaphysics of mind*, ed. John Haugland (Cambridge, MA: Harvard University Press, 1998), 207–40; Andy Clark, "Pressing the Flesh: A Tension in the Study of the Embodied, Embedded Mind?" *Philosophy and Phenomenological Research* 76, no. 1 (2008): 37–59; Mark-Oliver Casper, *Social Enactivism: On Situating High-Level Cognitive States and Processes* (Berlin: De Gruyter, 2019), 13ff.

1.2 A Normative Turn from Functional States to Intersubjective Statuses

Modeling cognition with information processing is inseparable from the development of information theory. To provide a general model for describing communication between sender and receiver, Shannon's information theory, defining information in mathematical or purely syntactic terms.[56] The purpose of the purely syntactic definition is to model information exchange between senders and receivers in general, independently of whether they are electronic devices or living organisms. The information processing-based models portray a new image of human beings, as Floridi calls it, "inforgs", informationally embodied organisms that are interconnected and embedded in the information environment.[57] Although modeling based on pure syntax was the original intention of information theory, information technologies theorists have been searching for suitable concepts that can add a semantic dimension to information. Semantic dimension can provide prescriptive links between information that are comprehensible to information processing agents. Attention has been focused on how the semantic dimension of information can be integrated into its definition so that it can be interpreted as syntactically sound data and meaningful data.[58]

Turning the concept of information from purely syntactically descriptive to semantically prescriptive is not only a theoretical issue, but also practice-oriented. The digitalization of everyday practices generates huge volumes of data. The word data comes from the plural of the Latin word "datum", the past participle of dare, meaning "thing given" or "a fact given or granted".[59] The use of the word data implies endorsement of it as evidence about physical reality or as directly received information. Data can thus be understood both as the content that is given about physical reality and as representing it in a structured form. The digitalization process does not just present reality in binary codes but incorporates social practices into a global process of data circulation that connects users, devices, and things, depending on the actual technical infrastructure, network protocols and their standardization. The data generated in digital social practices are clustered, structured,

---

[56] Shannon notes that from an engineering standpoint that information theory should not deal with semantics, as he writes: "semantic aspects of communication are irrelevant to the engineering problem. The significant aspect is that the actual message is one selected from a set of possible messages." Claude Shannon, "The Mathematical Theory of Communication," *The Bell System Technical Journal* 27 (1948): 379.

[57] Luciano Floridi, *The 4th Revolution: How the inforsphere is Reshaping Human Reality* (Oxford: Oxford University Press: 2014), 94.

[58] Luciano Floridi, *The Philosophy of Information* (Oxford: Oxford University Press, 2011), 92.

[59] Douglas, "Etymology of data." The term data was used first in mathematics to express "a fact given as the basis for calculation in mathematical problems" and was introduced into computer science in 1946 to present "transmittable and storable information by which computer operations are performed".

and extracted constantly into facts and their relations available to information processing agents. Against this background, information exchange cannot simply be modelled as a form-fixed exchange between ideal senders and receivers, but must consider the standardisation process of the transfer medium and the experience of what makes information processors ideal senders or receivers.

The consideration of facilitating information processors to maximize their access to information resources has promoted the growth of the Semantic Web and semantic technologies. Semantic Web was conceived as a way to make resources in the internet not only human-readable, but machine-readable as well.[60] The central tools of semantic technologies offer a range of data structuring methods that are completely different from classical formal data structures. Data can be semi-structured by these tools so that they can be combined into interrelated data and metadata. Yet, their relations are not fixed, but are assigned during their generation and use. As a result, networked intelligent applications such as smart assistants can imitate the use of human language by engaging in discursive practices with humans. Here, the model of syntactic programs processing the given data is no longer adequate to account for data environments supported by semantic technologies and for the information processors therein being formed for appropriate functions. The contemporary data environment, also known as cyberspace, is both a technological and a normative space that shared by human and artificial information processing agents. Chapter 3 will expand on the argument that cyberspace is not the opposite of reality, but rather an extension of social practices and materializes the norm-making processes therein. As the central argument of this book, it will explain how such a space can provide the conditions for information processing agents to act on normative grounds and become reciprocally authorized rational agents. To account for the role of cyberspace as a normative space for affording rational action, it is necessary to extend the computational functionalist model from a rationalist and normative perspective.

As outlined above, the notable difficulties of computational functionalism are how to explain how semantics is assigned to syntactic structures and how to relate mental representations to the physical world. Chapter 4 will introduce Davidson's truth-conditional approach to semantics and discuss how his externalized epistemology of empirical knowledge can help to resolve these difficulties. His theory of meaning attempts to provide a universal method for interpreting the meanings of expressions in a natural language, that is, by determining their truth conditions. Such a formal semantic approach reveals a recursive structure in language use, namely, interpreting the truth conditions of an object language from the metalinguistic level. This approach can explicate when an expression makes sense, that is, the conditions under which the expression can be taken to be true. And holding the belief of something to be true is considered fundamental to using language in a meaningful way. This theory of semantics expands the understanding of meaning

---

[60] Tim Berners-Lee and Mark Fischetti, *Weaving the Web: The Original Design and Ultimate Destiny of the World Wide Web by Its Inventor* (New York: Harper, 2000), 157f.

into the intersubjective interpretation in language practice. The origins of interpretable meaning move from subjective representations to the interaction between subjects in their shared environment. Accordingly, Davidson proposes triadic interaction as the basic model for generating thinking and judging whether others think, and he employs it to improve the Turing test. The corresponding triangulation is a method he proposes for agents to acquire knowledge about the world, others, and self in interaction.

The interpretive theory of meaning and the externalized epistemic account of knowledge touch on a broader question of how to ensure that the knowledge acquired by interacting agents is both interpretable and factual. This question involves the discussion of ontological relativity and ontological commitment by Quine and will also be discussed in Chap. 4. Clarifying ontological commitments in language use helps to understand why different agents in interaction are not only sharing the same environment, but also undertake the commitment to a shared world of meaning. Ontological commitment guarantees relations between objects that can be expressed and manipulated. The basic move that offers the possibility to undertake such commitments is to generate information patterns that can be recognized for interpretation. The interaction of multiple agents in a shared environment is the condition for the formation of patterns that can be recognized and interpreted by those agents.

To adequately describe the changing states of interactants in shared practices, a conceptual framework that begins not with individuals but with discursive interactions is needed. Chapter 5 will elaborate on Brandom's normative inferentialist approach, which provides a normative conceptual framework for explaining how discursive practitioners establish meaningful expressions and attribute mental states to each other, where mental states are replaced with normative statuses like entitlement and commitment. This normative conceptual framework is based on three theoretical foundations. The first is semantic inferentialism, which means understanding semantic content according to the role it plays in inference. The second is normative pragmatics, which is to place mental states such as beliefs in the normative space of reasons, where their implicitly committed normative status can be clarified through the game of giving and asking for reasons. The third is logical expressivism, that is, broadly defined logical vocabularies, including conditionals, modals, as well as normative and evaluative terms, have the role of expressing implicit practical procedures in a law-like form. The linguistic practice performed with expressive symbolic structures is a unique ability of humans as rational beings. According to the normative inferentialist view, it is the ability that enables humans to move from the natural world into the normative realm and to reciprocally license each other as rational agents.

The normative realm arising from shared practice is what Hegel calls "Geist", where linguistic practitioners recognize and authorize each other as members of discursive practice communities. Key to this insight is taking the use of logical language in discursive practices as the basis for authorizing and attributing mental states. Practicing discursive practices thus plays a constructive and normative role in the mutual attribution of mental states between individuals. According to this view, intrinsic intentionality, which is typically taken as the mark of the mental, can only be attributed to communities of discursive practices of reciprocal authorization. To practice the use of logical language is to take on a rationalist discursive commitment. The one who performs the inferential articulation in discursive practices is the agent of rational agency who is entitled to take on rationalist discursive commitments. This normative inferentialist approach to rational agents can be fruitfully applied to attribute and license the normative status of non-human rational agents that share the normative space of reasons with humans.

Why can a normative inferentialist understanding help to address the difficulties of computational functionalism? In this understanding, mental states are characterized as normative statuses in discursive practices. This normative turn breaks down the presupposed boundaries between internal states and the external world. The space of reasons constructed by discursive practices is not physical or psychological, but a normative space of mutually licensed deontic statuses. The normative space of reasons can serve as a more fundamental space for knowledge justification than the psychological or physical space. And cyberspace as a materialized normative space can bring the implicit practical norms and discursive commitments of shared practices into clear view with the help of the normative conceptual framework. This, then, is also the condition under which artificial intelligent agents with access to cyberspace resources can be recognized and authorized as rational agents. In particular, interpretable patterns of information formed in shared practices are semi-structured into available resources in cyberspace through semantic technologies. These data resources, which are constantly generated and updated, afford the possibility for human and artificial agents to act rationally as data processing agents that take on discursive commitments.

Chapter 6 will specify the role of the internet in the development of general-purpose artificial intelligence, taking the example of large language models represented by the GPT family. It will introduce how web resources have enhanced the GPT family. Internet-based large language models are like snapshots of the internet reflecting the conventions and diversity of everyday social practices. They not only represent a shared reality, but also express norms implicit in digital social practices. I will argue that in addition to representing digitized social practices, large language models can also be trained to acquire the expressive power of language. Artificial agents powered by expressive power in discursive practices can be recognized as explainable rational agents. On training artificial agents to be expressive I will make two proposals. One is that more diverse human users should be involved in interactive machine learning, so that different human preferences can be learned and thereby harmful biases and stereotypes can be reduced. Another is that when human users use the web and interact with intelligent agents, well-organized information management and reasoning structures, such as labeling information sources, can provide reliable learning material for artificial agents.

Finally, I would like to stress that the normative turn in understanding the mind has implications not only for the construction of normative technological environments and artificial agents, but also for understanding ourselves. Making implicit norms explicit can shed light on these implicit biases and can thereby help to pave the way for the enactment of unbiased norms of practice communities. A normative turn would orient technological agents to be centered on human social practices, rather than anthropomorphize them as individual persons. For it should become clear that it is we who authorize each other in common practice to make norms and afford the possibility for rational action by information processing agents.

Chapter 7 as a concluding chapter will summarise discussions in the previous chapters into three main points. First, cyberspace is the logical space that is being materialized, that is, it is an extension of the norm-making process of social practice. Cyberspace, supported by semantic technologies, can provide reasoning resources for the information processing agents that access it. Second, it is necessary to continuously form machine-readable and recognizable information patterns in cyberspace so that artificial information processing agents can share the realities of human practices. Third, trustworthy and interpretable artificial agents should be oriented in line with rational agents that are responsive to discursive practices. This orientation requires artificial intelligence technologies to be treated not just as the product of labs and development teams but needs to be actively regulated and shaped by social agencies. The concept of artificial intelligence also changes from an anthropomorphic orientation to being centered on human social practices. Accordingly, the question of whether machines can think can be answered in yes, machines can think, provided they take on rationalist commitments in the normative space of reasons shared with us. The Turing test will be determined in digital social practices, not in development labs alone.

Bibliography

Berners-Lee T, Fischetti M (2000) Weaving the web: the original design and ultimate destiny of the world wide web by its inventor. Harper, New York

Boole G (1854) An investigation of the laws of thought, on which are founded mathematical theories of logic and probabilities. Dover, New York

Brockhaus RR (1991) Realism and psychologism in 19th century logic. Philos Phenomenol Res 51(3):493–524

Casper M-O (2019) Social enactivism: on situating high-level cognitive states and processes. De Gruyter, Berlin

Cassirer E (1944) An essay on man: an introduction to a philosophy of human culture. Yale University Press, New Haven

Church A (1936) An unsolvable problem of elementary number theory. Am J Math 58(2):345–363

Clark A (2008) Pressing the flesh: a tension in the study of the embodied, embedded mind? Philos Phenomenol Res 76(1):37–59

Copeland J (ed) (2004) The essential Turing: seminal writings in computing, logic, philosophy, artificial intelligence, and artificial life: plus the secrets of enigma. Clarendon, Oxford

Davis M (1958) Computability and unsolvability. McGraw-Hill, New York

Douglas H (2022) Etymology of data. In: Online etymology dictionary. https://www.etymonline.com/word/data. Accessed 14 July 2022

Evans CL (2018) Broad band: the untold story of the women who made the Internet. Penguin Putnam, New York
Floridi L (2011) The philosophy of information. Oxford University Press, Oxford
Floridi L (2014) The 4th revolution: how the infosphere is reshaping human reality. Oxford University Press, Oxford
Fodor JA (1975) The language of thought. Harvard University Press, Cambridge
Frege G (1960) The foundations of arithmetic: a logico-mathematical enquiry into the concept of number (trans: Austin JL). Harper Torchbooks, New York
Frege G (1979) Boole's logical Culculus and the Begriffsschrift. [1880/1881]. In: Hermes H, Kambartel F, Kaulbach F (eds) Gottlob Frege: Posthumous writings. University of Chicago Press, Chicago, pp 9–21
Frege G (1997a) Begriffsschrift (1879): selections. In: Beaney M (ed) The Frege reader. Blackwell, Oxford, pp 47–79
Frege G (1997b) Thought (1918). In: Beaney M (ed) The Frege reader. Blackwell, Oxford, pp 325–346
Gödel K (1931) Über formal unentscheidbare Sätze der Principia Mathematica und verwandter Systeme I. Monatshefte für Mathematik 38(1):173–198
Gödel K (1986) On undecidable propositions of formal mathematical systems (1934). In: Feferman S, Dawson JW, Kleene SC, Moore GH, Solovay RM, van Heijenoort J (eds) Reprinted in Collected works. I: publications 1929–1936. Oxford University Press, Oxford, pp 338–371
Haugeland J (1998) Mind embodied and embedded. In: Haugland J (ed) Having thought: essays in the metaphysics of mind. Harvard University Press, Cambridge, pp 207–240
Hilbert D (1900) Mathematische Probleme. In: *Göttinger Nachrichten*, 253–297, and in Archiv der Mathematik und Physik, (3) 1 (1901), 44–63 and 213–237. Published in English translation by Mary Winton Newson, In Bulletin of the American Mathematical Society 8 (1902), 437–479
Leibniz GW (1951a) On the method of universality [1674]. In: Wiener PP (ed) Leibniz: selections. Charles Scribner's Sons, New York, pp 3–4
Leibniz GW (1951b) Preface to the general science [1677]. In: Wiener PP (ed) Leibniz: selections. Charles Scribner's Sons, New York, pp 12–17
Leibniz GW (1989a) Dissertation on the art of combinations [1666] (selections). In: Loemker LE (ed) Philosophical papers and letters: a selection. Kluwer Academic Publishers, Dordrecht, pp 73–84. https://doi.org/10.1007/978-94-010-1426-7
Leibniz GW (1989b) Letter to Nicolas Remond, [1714-1715]. In: Loemker LE (ed) Philosophical papers and letters. Kluwer Academic Publishers, Dordrecht, p 654
Leibniz GW (1989c) On the general characteristic, [ca. 1679]. In: Loemker LE (ed) Philosophical papers and letters. Kluwer Academic Publishers, Dordrecht, pp 221–228
Lieto A (2021) Cognitive design for artificial mind. Taylor & Francis, Milton Park
Mainzer K (2014) Die Berechnung der Welt. Von der Weltformel zu Big Data. C.H. Beck, München
Piccinini G (2004) Functionalism, computationalism, and mental states. Stud Hist Philos Sci A 35(4):811–833
Putnam H (1967) The nature of mental states. Originally published as "Psychological predicates". In: Captain WH, Merrill DD (eds) Art, mind and religion. University of Pittsburgh Press, Pittsburgh, pp 37–48. Reprinted in Mind, language and reality, philosophical papers 2. 429–440. Cambridge University Press, Cambridge, 1975
Russell B (1908) Mathematical logic as based on the theory of types. Am J Math 30(3):222–262
Shannon C (1948) The mathematical theory of communication. Bell Syst Tech J 27:379–423, 623–656
Suber P (1988) What is software? J Specul Philos 2(2):89–119
Tarski A (1936) The concept of truth in formalized languages. In: Tarski A (ed) Logic, semantics, metamathematics. Oxford University Press, Oxford, pp 152–278
Thagard P (2005) Mind: introduction to cognitive science. MIT Press, Cambridge

Turing AM (1937) On computable numbers, with an application to the Entscheidungsproblem. Proc Lond Math Soc s2-42(1):230–265

Turing AM (1950) Computing machinery and intelligence. Mind 59(236):433–460

Turing AM (1969) Intelligent machinery. In: Meltzer B, Michie D (eds) Machine intelligence 5. Edinburgh University Press, Edinburgh, pp 3–23. This paper was submitted as a report to the National Physical Laboratory in 1948 but had not been published

Chapter 2
How Minds Relate to Machines

How can machines and minds be related? The analogy between machines and minds has profoundly influenced cognitive science and AI research. Before delving into how machine models can be used to simulate mind in the field of AI research, it is important to clarify typical category mistakes in understanding minds. Clarifying category mistakes about what the mind is helps to orientate the practical goals of AI research and development. After that, this chapter will explore whether and how these category mistakes could be dissolved by the main approaches in AI research. Analogies between minds and machines will be discussed here, in particular the analogy between neural and circuit systems in cybernetics, the early functional analogies between minds and Turing machines in symbolic approaches, and the connectionistic analogies between brain plasticity and the learning capacity of artificial neural networks in sub-symbolic approaches. An examination of these approaches suggests that understanding minds in a naturalistic framework alone is insufficient for developing thinking machines. A normative framework is needed to articulate the conditions under which intelligence can be recognized and explained as taking place.

2.1 Category Mistakes

Mental conduct and bodily movement are usually treated as two things in everyday language use. This distinction makes it difficult to explain how humans act intelligently. Ryle indicates that this distinction is a category mistake in the understanding of mind and matter, which commits mental conduct to a false category.[1] A category mistake occurs when a concept is placed in the wrong logical category. A

[1] Gilbert Ryle, *The Concept of Mind*, 60th Anniversary ed. 1949 (Abingdon, Oxon; New York: Routledge, 2009), 6.

Y. Du, *Machines in Minds*, Studies in Applied Philosophy, Epistemology and Rational Ethics 78, https://doi.org/10.1007/978-3-032-15842-0_2

well-known example from Ryle is a person on the Oxford University campus, asking "Where is the university?".[2] The absurdity of this question is that the concept of university is not understood as a location or object such as campus, building, cafeteria, etc. As a concept referring to the overall function of these locations, it does not belong to the same logical category as them. Such mistakes are common in everyday language use. Semantic confusion revealed in the category of errors about mental conduct may contain ontological presuppositions about the mind.

2.1.1 The Mind-Body Problem in Spatial Metaphor

In his book *The Concept of Mind* (1949), Ryle begins by pointing out that the opinion that physical methods are unsuitable for explaining the mind is gradually becoming an "official doctrine".[3] According to this doctrine, mental states are immediately accessible to a person and their state changes are "self-evident", whereas physical changes need to be explained by the law of cause and effect. The occurrence and change of mental states cannot be explained by physical laws, as they have been put into a different explanatory framework. Ryle indicates that the official doctrine is accompanied by the "intellectualist legend" of "two lives" or "two worlds", i.e., the description of one's life is divided into two worlds, a private mental world consisting of events in the mind and a public physical world consisting of physical things.[4]

Nevertheless, it leads to difficulties of explanation, if one accepts the spatial metaphor in this doctrine but still tries to find a place for the mental in the physical world or tries to describe the physical in the same way as the mental. The explanatory gap comes from the presuppositions of this doctrine. On the one hand, it assumes that by merely observing the behavior of the material body, one can only access a physical description, which is inappropriate for describing the psyche. On the other hand, it assumes that a person has default "privileged access" to her own mental states.[5] This privileged access allows people to know themselves directly, but it also means that they cannot be known directly by others and can only indirectly know others' "inner worlds".[6]

The intellectualist legend is a legacy of "Cartesianism", as Ryle hints, it comes from "a more widespread anxiety about the implication of seventeenth century

[2] Ibid., 7.

[3] Ibid., 2.

[4] Ibid., 5.

[5] Ibid., 4. Yet Ryle stresses that, in fact, consciousness and self-awareness are not always transparent to people themselves. Not only is understanding other people indirect, but one's own thinking as comprehensible thoughts also requires mediums such as language. Particularly, he notes that the discovery of the subconscious suggests that a person cannot discover all her own mental events through introspective thinking.

[6] Ibid., 10.

mechanism".[7] This anxiety is accompanied by the mind-body problem that emerged under the influence of mechanics. The development of mechanics and machines provided an effective method of machine paradigm to study the motion of objects. In particular, the bodies of organisms were also understood as precision machines, and the way they move obey mechanical laws. This raises the question about the relationship between human thoughts and mechanical laws, and how the mind could relate to the body. Based on the powerful analogy between the body and the machine, either mechanical laws are equated with the laws of intelligent behavior, or there is a desire to distinguish human intellect from mechanical laws, mind from matter. In this way, the inquiry into mind becomes the dogma of the "ghost in the machine"- "the representation of a person as a ghost mysteriously ensconced in a machine".[8] This pictures the mind-body problem, which arises as soon as one presupposes that the mind and the body are two things that operate according to different laws, and then tries to find the principles by which they interact.

The search for the principles of mind-body interaction in this context is usually a search for a para-mechanical hypothesis of the functioning of the mind. The para-mechanical hypothesis holds that the mind actually operates according to rules similar to mechanical laws, so that minds are like special engines. Thereby the mind is classified into the same category as the body in the mechanical world, with the mind being some kind of substantial body-like existence, and changes in mental states similar to changes in physical states. As a result, psychological space is mistakenly regarded as the same category as physical space, and the myth of "two worlds" emerges. Those who make category mistakes usually seek some properties that may be shared by mental and physical, such as "things", "states", "processes", "effects", etc., in attempts to explain the interaction between the two worlds.[9] Ryle perceptively points out that because mechanics and the machine paradigm profoundly influence everyday language and thinking, this approach is actually describing the mind in mechanical terms:

> Since, according to the doctrine, minds belong to the same category as bodies and since bodies are rigidly governed by mechanical laws, it seemed to many theorist to follow that minds must be similarly governed by rigid non-mechanical laws. The physical world is a deterministic system, so the mental world must be a deterministic system. [...] *Responsibility, choice, merit* and *demerit* [...] have the congenial attribute of being only rather rigid.[10]

On the one hand, spatial metaphors are pervasively embedded in our everyday language, where the internal psychological space and the external physical space are typically separated. On the other hand, there is such inappropriate attempt of bringing the mental under mechanical law-like descriptions. Terms such as responsibility, evaluation, etc., which are supposed to describe intelligent behavior, in turn are analogized to mechanics and become inherent properties of things. The key to the

---

[7] Ibid., 12f.

[8] Ibid., 5, 8.

[9] Ibid., 9.

[10] Ibid., 10.

category mistake of the mind-body problem here is spatial metataphor, which presupposes two worlds within a mechanical explanatory framework.

2.1.2 The Category Mistake of Embodying Mind

The category mistake between mind and body occurs in presupposing them as different categories, yet explaining the mind by the very same mechanical laws that apply to the body. How, then, is it possible to explain mental conduct away from mechanical laws? Ryle notes that this requires placing not only the mind but also intelligent bodily activities in a logical rather than a mechanical explanatory framework. He argues that an "intellectualist doctrine" that merely attempts to free the mind from mechanism does not explain how the mind and body interact, because according to this doctrine one's intelligent behavior is always deduced from the relevant true propositions.[11] The reliance on the derivation of true propositions leads to an infinite regress problem. That is, since action is guided by some true propositions and these propositions need to be derived from some other propositions and so on, then an actor must perform a great deal of derivation before acting.[12] The deeper question involved here is how a true proposition can somehow initiate action.

To account for the relationship between knowledge and practice, Ryle distinguishes between two different forms of knowledge, "knowing that" and "knowing how". "Knowing that" is propositional knowledge, which constructs theories. By contrast, "knowing how" is procedural knowledge, which is the way to act. While knowing how includes knowing how to use propositional knowledge appropriately, it cannot be reduced to accepting propositional knowledge or deriving from it.[13] Just as describing a person as intelligent does not mean that the person has mastered some true proposition, but rather that the person knows how to do something. Ryle writes:

> What distinguishes sensible from silly operations is not their parentage but their procedure, and this holds no less for intellectual than for practical performances. 'Intelligent' cannot

[11] Ibid., 16. He sees this reverse as an attempt "to define intelligence in terms of the apprehension of truth, instead of the apprehension of truths in terms of intelligence".

[12] Ibid., 19f.

[13] Ibid., 20. There are two positions that deny the dichotomy between "knowing how" and "knowing that". Intellectualists argue that "knowing how" is a special kind of "knowing that", because "knowing that" can be understood as a broader ability. They hold that Ryle unfairly treats "knowing how" only as an ability and "knowing that" as a thinking relation. Cf. Jason Stanley and Timothy Williamson, "Knowing How," *The Journal of Philosophy* 98, no. 8 (2001): 411. On the contrary, radical anti-intellectualism considers "knowing that" to be a form of "knowing how" because knowing a proposition is acquiring its associated belief, which can be understood as a special kind of ability. Cf. Stephen C Hetherington, *How to Know: A Practicalist Conception of Knowledge* (Malden, MA: Wiley-Blackwell, 2011), 42. I will argue in the following that Ryle himself tends to consider that "knowing that" presupposes the ability to "know how" to use propositions, and not vice versa.

> be defined in terms of 'intellectual' or 'knowing how' in terms of 'knowing that'; 'thinking what I am doing? does not connote 'both thinking what to do and doing it'. When I do something intelligently, i.e. thinking what I am doing, I am doing one thing and not two. My performance has a special procedure or manner, not special antecedents.[14]

Procedural knowledge occurs in practical processes, which can be theorized as propositions, but is not derived from them. Being intelligent corresponding to "knowing how" underlies being intellectual corresponding to '"knowing that", not the other way round. The relation between intelligent actions and true propositions in the "intellectualist doctrine" should thus be reversed. Thus, "[u]nderstanding is a part of knowing *how*" because understanding a proposition can be treated as knowing how to take it as true and how to use it appropriately.[15] This means that mastering propositional knowledge as true beliefs requires knowing how to arrive at such beliefs.

Moreover, "knowing how" can be understood as a variety of practical skills, ranging from knowing how to swim to knowing how to play chess. Ryle points out that these skills are not causes that drives action, but rather capacities, habits, or bents in virtue of which action is performed.[16] Practical skills may involve training in sensorimotor bodily techniques, or the accumulation of individual experience. Yet whether it is bodily skill or tacit knowledge that has not yet been articulated in propositional form, "knowing how" differs from "knowing that" in that it is situational and adaptive, not categorical.[17] Mechanistic explanations that draw on causes and effects between categorical entities can hardly account for the machinery of such practical skills.

To adequately describe "knowing how", Ryle employs the term "disposition", which basically means potentiality, capability, and the tendency to do something.[18] Considering the realization of skills, procedural knowledge is not an inherent ability or habit, but latent competence that needs to be actualized. The term disposition can characterize the complex relationship between skill and exercise, potential and actualization, because dispositions can be expressed in sentences with "if, then" structure, stating the conditions under which the action will occur. In most cases, such conditionals are implicit in the dispositional concept. For example, the glass is

[14] Ryle, *The Concept of Mind,* 20.

[15] Ibid., 41.

[16] Ibid., 33. He indicates that "our inquiry is not into causes (and a fortiori not into occult causes), but into capacities, skills, habits, liabilities and bents."

[17] The term tacit knowledge is coined by Ponalyi, who argues that what can be expressed is only part of the tacit knowledge that arises from experience. He considers that animals possess tacit knowledge as well, and that humans have the advantage of being able to express it and therefore transmit it. The fact that humans always know more than they can tell is also known as Ponalyi's paradox. Michael Polanyi and Sen, Amartya. *The tacit dimension* (Chicago: University of Chicago Press, 2009), 1–26. Nielsen notes that because Dreyfus and Dreyfus replaced tacit knowledge with the term intuition, tacit knowledge is sometimes misunderstood as theoretically inexpressible. Cf. Klaus Nielsen, "The Concept of Tacit Knowledge—A Critique," *Outlines. Critical Practice Studies*, 4 (2) (2002): 6ff.

[18] Ryle, *The Concept of Mind*, 107.

fragile means that if the glass falls to the ground, then it might break. Or he is a sentimental person means that if he watches a sad movie, then he might shed tears. Yet the complexity of the conditions required for actualizing various dispositions varies. Ryle distinguishes between simple and complex dispositions. He treats physical properties and behavioral habits as "single-track" dispositions, and the conditions for actualizing simple dispositions are fixed.[19] In contrast, intelligent behaviors contain "multi-track" dispositions and are not merely reflexes. The conditions for achieving these behaviors are very broad, because the behavior is not an automatic response to stimuli, but involves responding in different ways, and practicing how to respond in different ways.

"Knowing how" as a complex of dispositions can be developed through exercise.[20] Someone who knows how to swim well has both a floatable body and swimming skills acquired through exercise. Both acquired practice and a suitable environment are conditions for performing intelligent behaviors that incorporate complex dispositions. McGeer points out that the key message of procedural knowledge is that "in exercising an intelligent capacity [...] the agent will be producing a pattern of flexible, adaptive, creative behaviour over time, as she tunes and retunes her activities in continuing response to the world around her."[21] In other words, performing intelligent behavior is not just habitually replicating acquired behavioral patterns, but involves a learning process of adapting and modifying behavioral patterns in response to environmental feedback. It therefore also shows that agents capable of acting intelligently are formed in practice by exercising skills.[22] Complex dispositions are not statically fixed within the agent, but rather are trajectories constituted by the interaction of the agent and the environment, depending on how the agent would behave and how dispositions could be actualized. This account breaks down the opposition between mind and body, the internal world and the external environment.

It is clearly inappropriate to think of the mind as some kind of body-like entity and look for the principles by which it interacts with a mechanical body. Having presupposed given entities and principles of their interaction in a mechanistic view,

[19] Ibid., 31–32.

[20] Ryle writes: "Knowing how, then, is a disposition, but not a single-track disposition like a reflex or a habit. Its exercises are observances of rules or canons or the applications of criteria, but they are not tandem operations of theoretically avowing maxims and then putting them into practice. Further, its exercises can be overt or covert, deeds performed or deeds imagined, words spoken aloud or words heard in one's head, pictures painted on canvas or pictures in the mind's eye. Or they can be amalgamations of the two." (Ibid., 34)

[21] Victoria McGeer, "XV-Intelligent Capacities," *Proceedings of the Aristotelian Society* 118 (3) (2018):365. In particular, she notes that the intrinsic features that constitute the dynamic performance of intelligent capacities are not constant but can be modified in response to feedback.

[22] According to McGeer, dispositions can be divided into passive object-centered, and active agent-centered. The former is physical in nature, while the latter involves acquired disposition. Both habits and intelligent abilities fall into the latter category, and Ryle distinguishes between these two types of acquired dispositions by treating the execution of procedural knowledge as practicing intelligent capacity. Ibid., 349.

addressing the mind-body problem would fall into the category mistake. It is necessary to express intelligent activity in a way that is different from the mechanistic explanations of motion.

2.1.3 *The Category Mistake of Encoding Mind*

On how to appropriately express intelligent activity, Ryle proposes dispositional propositions rather than categorical propositions. Rather than simply taking categorical propositions to be true, mental conduct involves making testable hypothetical propositions in the modal sense of "could" and "would", as Ryle writes, "roughly, the mind is not the topic of sets of untestable categorical propositions, but the topic of sets of testable hypothetical and semi-hypothetical propositions".[23] Dispositional propositions are not reports about inner states or fixed labels that cannot be modified, but are semi-hypothetical propositions that can be tested and adjusted.[24] The mind is accordingly characterized as the disposition to make hypotheses that can be expressed in the form of dispositional propositions. In other words, the mind is not certain sort of category, but the disposition and power to categorise.

Dispositional propositions account for the complex relation between knowledge and practice. Thinking with propositional knowledge is a form of practical skill, namely knowing how to use testable semi-hypothetical structure to express and guide practice. In other words, thinking and doing are both practices. Thinking in the form of theorizing is a discursive practice of exploring rules drawing on linguistic tools, rather than mechanically following given intellectual "laws of thought". Specifically, semi-hypothetical propositions can be used as linguistic tools to compose counterfactual subjunctive conditionals. This allows semi-hypothetical statements to form theories that can be revised. Interpretations of how intelligent behavior operates are thus not reducible to causal effects. In particular, the semi-hypothetical structure of dispositional propositions allows for a trial-and-error "as if" attitude in understanding. By means of this pragmatic structure, the relationship between complex dispositions and their actualisation conditions can be flexibly articulated. By clarifying that intellectual thinking comes from practice, theorizing is no longer the apprehension of some universal law but the construction of laws through semi-hypothetical propositions. Ryle uses the path metaphor to illustrate the process of theory building, in which the paving stone is the dispositional proposition, and the theorizing is both path-making and path-using.

First, theorizing is not about making paths according to given laws, but rather starting with the available paving stones. He reminds that "we have to learn to use statement of particular matters of fact, before we can learn to use the law-statements

[23] Ryle, *The Concept of Mind*, 34.

[24] Ibid., 131.

which do or might apply to them".[25] Namely, the "laws" generalized by theorizing are statements about statements, such as relations, properties etc., no longer at the same level as direct statements. The legitimacy of such meta statements needs to be licensed, otherwise it will lead to a regress in theorizing. This requires a temporary license of the inferential operation as a path to the facts. He describes it as a kind of inference-ticket:

> A law is used as, so to speak, an inference-ticket (a season ticket) which licenses its possessors to move from asserting factual statements to asserting other factual statements. It also licenses them to provide explanations of given facts and to bring about desired states of affairs by manipulating what is found existing or happening.[26]

The metaphor of the inference-ticket is used to refute the traditional view of "inferential rails". This view assumes that there are established facts at both ends of the inference, and that reasoning is the discovery of the track that slides from one given fact to another. The problem, Ryle points out, is that in this reasoning process the "legitimacy of the inference is assumed all the time".[27] Yet reasoning cannot be reduced to following a pre-existent track to a given destination, otherwise it is a singletrack disposition, a mechanical rather than an intelligent disposition. The reasoning process should consist of a series of inferential operations, each of which involves licenses that can be challenged and revoked. The skill of licensing inferential operations is present in our social interactions in that "[inference-tickets] license us to predict, retrodict, explain and modify these actions, reactions and states".[28] The practice of licensing inferential operations in everyday discourse, is usually in the form of argumentation. This indicates the social and discursive nature of reasoning. The mere encoding of declarative knowledge can only record rather than simulate rational thinking. I will return to this inferentialism in Chap. 5 in the light of Brandoms approach.

Second, theory building is about paving paths that one can follow and can pass through repeatedly. Paths are repeatable reasonings from premises to conclusions. The process of theory building is both path-making and path-using, as Ryle puts it, "[…] the thinker is both using his near-mastery of his theory and still schooling himself to master it perfectly."[29] In this sense, theory building is sometimes a process of making paths, sometimes using paths made by others, and sometimes making paths afforded by other paths. Therefore, paths are somehow objective in the sense that they are universally available to any rational agent. Ryle argues that the

---

[25] Ibid., 105. Factual reports and laws of explanation belong to different levels of discourse, he argues, just as arithmetical statements and laws of arithmetic belong to different levels of discourse.

[26] Ibid. Brandom adopts and further develops this inferentialist account with a detailed analysis of the discursive nature of knowledge and the intersubjective origins of knowing. I will elaborate on Brandom's account in Chap. 5.

[27] Ibid., 106.

[28] Ibid., 108.

[29] Ibid., 264. Some of these paths will lead to the discovery of a destination, and some will not. He understands theory building in a broad sense, including scientific discoveries, in addition to everyday speculation, predictions, etc.

universal availability of paths has its roots in the public nature of languages use.[30] The meaning of an expression is thus shareable in its availability, but it is also related to the way it is used by individuals. He also emphasizes that language use is the linguistically structured act, such as the proper use of indicative sentences reflecting the act of making judgements, and the proper use of logical conjunctions reflecting the act of reasoning.[31] The act of thinking is thus not described by language, but is carried out through it. Such intelligent activity is not the mechanical manipulation of symbols according to given rules, but the enactment of norms for the operation of symbols by means of linguistic tools.

So far, two implicit assumptions of the dogma of the "ghost in the machine" have been revealed. One is the reversal of the relation between comprehension of propositional knowledge and knowing how to use dispositional propositions, and the other is the misconception that theorizing takes place primarily within the mind rather than in intersubjective discursive practices. This dogma treats bodily actions and reactions that obey mechanical laws as intelligent behavior on the one hand and mystifies minds as being subject to intellectual laws on the other. Ryle uses the concept of procedural knowledge to explain how practice and understanding are intertwined, treating the performance of intelligent capacities as exercising and learning complex dispositions, rather than moving mechanically while understanding intellectually. Instead of following given laws, intelligent performance treats procedural knowledge in practice as adaptable behavioral patterns, and its theorization as paths to thought. This reflection on the mind-body problem and insights into intelligence are useful for understanding the various analogies between machines and minds. The next subchapter will show how the dogma of "the ghost in the machine" is abandoned by AI research, and yet how it manifests itself in new forms.

2.2 Computer Analogies and Challenges

Although the mechanistic picture of the body and mind has been overtaken by more recent scientific discoveries and theories, its conceptual framework has been carried over into everyday discourse and has profoundly shaped how the mind is understood. Since Turing put forward the path of thinking machines, simulating intelligence through modern computer technologies became has become one or a class of research fields, where analogies between biological and circuit systems, between minds and programs, between perceptual plasticity and machine learning arise. I will elaborate on these analogies in this subchapter and clarify that they are based on a functional understanding of intelligence, where neither mind nor machine are taken to be object-like entities. However, AI research based on the functionalist framework of the mind-machine analogy may lead to the fallacy of begging the

[30] Ibid., 16.

[31] Ibid., 264.

question. For on the one hand AI research aims to mimic human intelligence, and on the other hand computer models of AI research are used in cognitive science to understand human cognition. A more fundamental theoretical framework of information processing that allows these analogies to hold needs to be expanded and explicated.

2.2.1 Cybernetic Machinery vs. Artificial Intelligence

As a research field, machine intelligence can be traced back to at least two important series of conferences, which laid down the basic concepts for studying and modeling intelligence from two different ways. On one side is the Massey Conference with cybernetics as the core theory, and on the other side is the series of conferences starting with a workshop at Dartmouth to ground the foundations of AI as a research discipline.

The Macy Conference refers to a series of conferences held between 1946 and 1953 with the initial goal of finding a "general science of the workings of the human mind".[32] The conferences were chaired by McCulloch, who with Pitts designed the first computational model to simulate a biological neuron in 1943.[33] Inspired by the machine modeling approach, the conferences encouraged the exploration of universal models that simulate how organisms organize and perform intelligent behavior. In the interdisciplinary effort to develop universal models, information theory became a central methodology. Particular attention was paid to the information theory proposed by Shannon. In his view, the basic problem of communication is how to replicate information with less loss, and that the engineering aspect of this problem is not related to semantics but has to do only with the form system that carries the information.[34] By introducing the concept of entropy as a measure of probabilistic uncertainty, Shannon defines information as a method of calculating entropy for reducing uncertainty in the process of communication.[35] This theory provides a purely syntactic way to define information so that information theory can be applied to different kinds of physical systems. In this way, actual information processing in mechatronic systems can be analogous to information processing in biological systems, and different systems are distinguished from each other only by their complexity.

---

[32] Heinz von Foerster, *Cybernetics: Circular causal and feedback mechanisms in biological and social systems. Transactions of the seventh conference* (New York: Josiah Macy Jr. Foundation, 1951), vii.

[33] Other participants included Ross Ashby, Heinz von Foerster, Margaret Mead, John von Neumann, and Norbert Wiener, etc. Summaries of each conference can be seen in American Society for Cybernetics, *Summary: The Macy Conference.* https://asc-cybernetics.org/foundations/history/MacySummary.html.

[34] Shannon, "The Mathematical Theory of Communication," 379.

[35] Ibid., 415ff.

Since 1950, cybernetics gradually became a dominant interdisciplinary theory in this series of conferences, and the conference title was also changed to "Cybernetics. Circular, Causal, and Feedback Mechanisms in Biological and Social Systems".[36] Cybernetics was introduced by Wiener in his book *Cybernetics: or Control and Communication in the Animal and the Machine* (1948). Derived from its original meaning of steering and governing, the term cybernetics is defined here as a purposive feedback mechanism.[37] Wiener regards Leibniz as the pioneer of cybernetics for his conception of the universal calculus to reason about knowledge in different domains.[38] This conception is manifested in cybernetics as a circular causal process of information feedback in a system. On this basis, cybernetics aims to develop an interdisciplinary approach to the study of systems, including biological, social, and technological ones. This approach uses circular causal models of information processing to characterize intelligent operations of systems, such as coordination and regulation.[39] In defining intelligent systems by combining information theory and systems theory, new terms such as feedback loop and circular causality are introduced as general features of intelligent operations of either biological, social, or technological systems. As a result, analogizing biological nervous systems to the architecture of electronic computers became a trend at the time.

The other important conference series is represented by the Dartmouth Summer Research Project on Artificial Intelligence in 1956, which is often considered to be the starting point of the field of AI research.[40] In its research proposal drafted in 1955, this new research field was conceived as a machine approach to simulate various aspects of intelligence:

> The Study is to proceed on the basis of the conjecture that every aspect of learning or any other feature of intelligence can in principle be so precisely described that a machine can be made to simulate it. An attempt will be made to find how to make machines use language, form abstractions and concepts, solve kinds of problems now reserved for humans, and improve themselves.[41]

---

[36] Under the central theme of information theory and systems theory, the topics of the conferences changed annually. The first conference was titled "Feedback Mechanisms and Circular Causal Systems in Biological and Social Systems" and the second became "Teleological Mechanisms and Circular Causal Systems". *American Society for Cybernetics*, Summary: The Macy Conference. https://asc-cybernetics.org/foundations/history/MacySummary.html.

[37] Nobert Wiener, *Cybernetics: Or Control and Communication in the Animal and the Machine* (Cambridge, MA: MIT Press, 1985), 11–12.

[38] Ibid., 12.

[39] W. Ross Ashby, *An Introduction to Cybernetics* (London: Chapman & Hall, 1957), 1–3.

[40] Main attendees of the conference are John McCarthy, Claude Shannon, Marvin Minsky, Arthur Samuel, Trenchard Moore, Ray Solomonoff, Olive Selfridge, Allen Newell, and Herbert Simon. Their motives were different. Shannon wanted to apply information theory to computing machines and brain models, while McCarthy and Minsky wanted to build a new research field. Newell and Simon revealed a Logic Theorist program. Rochester was interested in networks of neuron-like elements. Cf. Nils J. Nilsson, *The Quest for Artificial Intelligence: A History of Ideas and Achievements* (Cambridge: Cambridge University Press, 2009), 79–81.

[41] John McCarthy, Marvin Minsky, Nathaniel Rochester, and Claude Shannon, "A Proposal for the Dartmouth Summer Research Project on Artificial Intelligence, August 31, 1955". *AI Magazine*

Since the aim is to implement machines that mimic intelligent behaviors and manners, seven specific goals for intelligent machines are proposed: "(1) Automatic computers; (2) How can a computer be programmed to use language; (3) Neuron nets; (4) Theory of the size of calculation; (5) Self-improvement; (6) Abstraction; (7) Randomness and creativity".[42] The diversity of methods plays a significant role in the exploration and development of the new field. The term "Artificial Intelligence" was coined by McCarthy and defined the discipline and engineering of making intelligent machines.[43] In contrast to cybernetics, AI research began by focusing on how to simulate different intelligent behaviors with digital computers, rather than exploring a universal intelligent system. Intelligence is here an engineering goal of what digital computers can do.

Remarkably, the field of AI research drives and contributes to the emergence and development of cognitive science. A few months after Dartmouth conference, a symposium on information theory was held at MIT. In addition to some participants from the Dartmouth conference attending the meeting, there were also researchers from other disciplines like Chomsky, who was working on a mechanical translation project at that time. Chomsky discussed his paper *Three Models for the Description of Language*, in which he describes the linguistic competence to produce sentences in terms of a symbolic manipulation approach.[44] Many other important works were published during this period and the interdisciplinary movement to study cognition gradually expanded, until the term "cognitive science" was first introduced in 1973.[45] Cognitive science, as an interdisciplinary paradigm, includes disciplines such as philosophy, anthropology, biology, linguistics, psychology, neuroscience, where AI approaches are primarily used as tools to study and model cognitive processes. Because information technologies and computer simulation are used as core methods for modeling, especially to modeling information processing, this research approach is also known as the "information-processing paradigm" of cognition.[46] As Dennett points out, AI research brings constructive insights to cognitive science and philosophy of mind on the one hand, and AI becomes a philosophical artifact on

27, no. 4 (2006): 12.

[42] Ibid., 12–14.

[43] Nilsson, *The Quest for Artificial Intelligence,* 78. According to Nilsson, the reason for using the term "Artificial Intelligence" is to mark a distinction from previous approaches. One is to distinguish it from cybernetics and move away from a focus on analog feedback, and the other is to distinguish it from Shannon and McCarthy's previous paper collection *Automata Studies*, which dealt primarily with the theory of automata.

[44] Noam Chomsky, "Three models for the description of language," *IRE Trans. Inf. Theory* 2 (1956): 113–124.

[45] Cf. Nilsson, *The Quest for Artificial Intelligence,* 40, 91. In the cognitive turn, important works includes Miller's *The Magic Number Seven, Plus or Minus Two* (1956), Chomsky's Review of B.F. Skinner's Verbal Behavior (1959), Selfridge and Neisser's *Pattern Recognition by Machine* (1960), and a series seminal works in AI research.

[46] Ibid., 73.

the other.[47] With the help of AI approaches, the philosophical question of how knowledge is possible can be explored in an engineering way. AI, as a modeling tool and as a model to be built, contributes both to theorizing cognition from engineering practice and to testing theories about cognition.

Cybernetics and AI research are significantly divergent in their theoretical orientations and approaches. First, in terms of the research object, cybernetics seeks models that characterize intelligent systems in a unified way, whereas AI research starts with realizable intelligent behavior. Cybernetics is concerned with feedback models that can be generally applied to automated mechanical systems, biological systems, or social systems, modeled in a self-regulated or self-organized manner.[48] In contrast, AI research focuses on how to implement specific intelligent behaviors, such as logical reasoning, natural language processing, learning, etc. Second, in modeling information processing, cybernetics takes a behaviorist approach to how information processing takes place within an intelligent system, whereas the cognitivist approach attempts to move away from behaviorism. Inspired by the study of artificial neurons, cybernetics models information processing in sensorimotor systems as feedback loop patterns in forms of sensory memory or body memory.[49] Cognitivists argue that such behaviorist style cybernetic models of communication simplify the higher functions of the entire person in processing information, or treat it as a black box, a model of input-output relations.[50] Yet in cybernetics, the black box method can be used to identify unknown systems capable of processing information properly by testing repeatable input and output relations.[51] Comparing the two methods, how a system outputs intelligent behavior is opaque in cybernetics, whereas cognitivist AI research conjectures and constructs rules of intelligent behavior.[52] The distinctions between cybernetics and AI research in the cognitive turn also underlie the debate between symbolic and sub-symbolic approaches to AI.

---

[47] He claims that "AI is better viewed as sharing with traditional epistemology the status of being a most general, most abstract asking of the top-down question: how is knowledge possible? Daniel C. Dennett, "AI as Philosophy and as Psychology," in *Philosophical Perspectives on Artificial Intelligence*, ed. Martin Ringle (New York: Humanities Press, 1978), 60.

[48] Early cyberneticists were more concerned with modeling biological systems. Later second-order cyberneticists realized that understanding systems requires multi-order recursion and second-order observers. Therefore, automated systems modeling should not only be self-organizing, but also self-producing. Second-order cybernetics is represented by Luhmann, Maturana and Varela, etc.

[49] Wiener, *Cybernetics: Or Control and Communication in the Animal and the Machine*, 122.

[50] As a term, the black box originally referred to a terminal element in an electronic circuit. This term was introduced by cybernetics and system theory to allow for a functional description of a device in terms of inputs and outputs, without knowledge of the internal workings. Cf. Mario Bunge, "A General Black Box Theory," *Philosophy of Science* 30, no. 4 (1963): 346–58.

[51] Wiener, *Cybernetics: Or Control and Communication in the Animal and the Machine*, xi.

[52] Chomsky relates cognitivism to rationalism, and behaviorism to empiricism. His argument is that behaviorists view sensory input as directly available sensory data, whereas cognitivists consider that there are inherent rules that regulate and shape perceptual experience and psychological processes. One premise of cognitivism is that humans have evolutionary or culturally based a priori knowledge of information processing that allows them to perceive and behave in particular ways. This means that even the most basic patterns of sensory information communication involve the

2.2.2 *Symbolic AI, the Computational Theory of Mind, and the Semantic Issue*

Symbolic AI refers broadly to the approaches that have dominated AI research since the 1960s through the 1990s, with many achievements from the early days of AI research, such as automatic theorem deduction, chess playing, and expert systems. These approaches share a common cognitivist assumption that symbolic manipulation systems are the essential manner to achieve intelligent behavior.[53] Symbolic AI approaches are usually based on well-structured databases with given rules to automate the reasoning process. Classical symbolic AI can perform tasks in a regional domain according to given rules, an execution process that is understandable and therefore known as "Good Old-Fashioned AI".[54]

A famous example is ELIZA, a chat program developed by Weizenbaum in 1966 that simulates a psychotherapist and talks to human clients. Although it can be easily detected in the Turing test as not being a real person, Weizenbaum notes that during the test, many people may still attribute traits of human behavior such as emotional and social factors to the computer's outputs. This tendency of human users is also known as the ELIZA effect, a tendency to unconsciously anthropomorphize the computer in their interactions with it.[55] The ELIZA effect shows, on the one hand, symbolic AI can convincingly simulate human intelligent behavior in certain domains, and on the other hand, it raises the question of whether it should simulate all aspects of human intelligence. Weizenbaum critically states in his book *Computer Power and Human Reason: From Judgment to Calculation* (1976) that "there are some acts of thought that *ought* to be attempted only by humans", and in particular that work regarding values should only be done by humans.[56] He argues that computers should at best be used as a tool or "powerful metaphor for helping us to understand many aspects of the world" at most, rather than being misused to

behavioral models and conceptual frameworks that prescribe such ways of perceiving. Cf. Noam Chomsky, *Empiricism and Rationalism. Excerpted from Language and Responsibility* (New York: Pantheon, 1977).

[53] As proposed by Newell and Simon, the "Physical Symbol System Hypothesis" postulates that "intelligent beings are physical symbol systems". This hypothesis assumes that symbol systems can be implemented as physical systems or hardware for symbol manipulation, and that symbols and their composition are the internal structures of the physical system. That is, the physical system is the materialization of symbolic structuring during symbolic manipulations. Antonio Lieto, *Cognitive Design for Artificial Mind* (Milton Park: Taylor & Francis, 2021), 10ff.

[54] Haugland defines Good Old-Fashioned AI as AI that performs rational operations on symbols. John Haugland, *Artificial Intelligence. The very Idea*, 6th Printing (Cambridge, MA: MIT Press, 1993), 228.

[55] Douglas Hofstadter, "The Ineradicable Eliza Effect and Its Dangers," In *Fluid Concepts & Creative Analogies: Computer Models of the Fundamental Mechanisms of Thought* (New York: Basic Books, 1995), 157.

[56] Joseph Weizenbaum, *Computer Power and Human Reason: From Judgment to Calculation* (New York: W. H. Freeman and Company, 1976), 12f.

simulate all aspects of human beings.[57] Weizenbaum's considerations both explore the ethical issues of AI as an automated tool for human labor and question the anthropomorphic image of AI.

A comparable criticism of AI development can be seen in Dreyfus' *Alchemy and Artificial Intelligence* (1965), in which Dreyfus argues that it is impractical to make machines that can think like humans the ultimate goal of AI development.[58] He outlines the areas of AI development at the time, such as games, pattern recognition, and language processing, and points out that limiting human capabilities to these areas is not sufficient to develop human-like AI. In other words, he argues that human intelligence is not limited to computational ability. Dreyfus holds that not all mental processes can be formalized as information processing, because human perception and consciousness are ambiguous rather than discrete and determinate.[59] This criticism points out that symbolic AI approaches rely on clear structures of data and given rules of reasoning, which cannot accomplish the goal of fully simulating the human mind.

In fact, these criticisms address the controversial assumption behind symbolic AI, namely the functionalist computational theory of mind, which assumes the mind to be functionally a computational system. This view can be traced back to McCulloch and Pitts' 1943 article providing a logical model for neural networks, in which Turing machines were thought to be capable of offering a model for neural activity.[60] This view gradually developed into the assumption that all human cognitive processes can be simulated by machine models. In this context, the term simulation takes on a methodological significance that goes beyond imitation. Since cognitive states are understood here in terms of their functional states, which need not be reduced to biological states but are assumed to be realisable by multiple physical systems, computational models that simulate human intelligence do not merely mimic it but they are functionally intelligent when they carry out the same symbolic operations.[61]

The functionalist theory of mind, which was explicitly formulated by Putnam in 1967, states that the mind is just the functional way in which physical systems are organized.[62] Putnam argues that mental states, such as pain, can be characterized as

---

[57] Ibid., 277.

[58] Hubert L. Dreyfus, *Alchemy and Artificial Intelligence* (Santa Monica: Rand Corporation paper, 1965), 10ff.

[59] Ibid., 66–84. Yet in terms of modeling non-discrete states by discrete machines, Turing had elaborated in the response section of his 1950 article on why discrete machines can be a useful modeling too. Dreyfus later focuses his arguments on intuition and tacit knowledge.

[60] Warren S. McCulloch and Walter Pitts, "A logical calculus of the ideas immanent in nervous activity," *The Bulletin of Mathematical Biophysics* 5, no. 4 (1943): 129ff.

[61] Hilary Putnam, "Psychological Predicates," in *Art, Mind and Religion*. eds. W. H. Captain and D. D. Merrill, 37–48 (Pittsburgh: University of Pittsburgh Press, 1967). Reprinted as "The Nature of Mental States," in *Mind, Language and Reality, Philosophical Papers*, Vol. 2, 429–40 (Cambridge: Cambridge University Press, 1975), 434f.

[62] Ibid., 435.

functional states that play a certain role in the "functional organization" of the organism.[63] Mental states, including those that do not directly trigger behaviour, such as values and beliefs, can be modeled as functional mechanisms, serving as mediators between input and output.[64] Since the mind is modeled as a machine model, mental states as functional states of the machine model do not need to correspond one-to-one with neurophysiological states. Accordingly, the same mental state is thought to be realizable by different physical states; e.g., organisms with different physiological structures can have "pain states" that play the same role.[65] In this way, the machine-state functionalist account refutes the mind-brain identity theory and emphasizes the irreducibility of the mind. Note that this view suggests that mental states are necessarily embodied and realized by matter, but not limited to a specific physical form.[66] Thus, on this view, it is possible for a silicon-based machine or a Martian to realize human mental states as well. This view provides a theoretical basis for the possibility to realize mental states by AI systems. Yet, how inputs and outputs are mediated through the mind as a machine model is yet to be further explored, especially the operational mechanisms that are treated as internal information processing.

In response, Fodor proposes the "language of thought" hypothesis. He critically adopts Chomsky's view that language competence derives from intrinsic faculties of language users, particularly given the productivity in natural language use and the fact that the mind can systematically make sense of meaning.[67] In this regard, Fodor argues that there is an intrinsic system of representations mediated by language that underlies computation.[68] Representations are understood as internal structures that carry information, resembling syntactic structures, with semantic properties. The function of mental states is to compose such mental representations, just as propositional attitudes arrange propositional contents.[69] The cognitive process is hence the computational process of composing mental representations into

---

[63] Ibid.

[64] Ibid. Putnam calls such universal machine model as a "Probabilistic Automaton", similar to the Bayesian brain hypothesis that will be elaborated in the next section.

[65] Ibid., 228.

[66] Hilary Putnam, *Representation and Reality* (Cambridge, MA: MIT Press, 1988), xii.

[67] The use of natural language is productive because "speakers—despite the fact that they are finite creatures, limited with respect to their experiences, memories, computational power, and so forth—can produce and understand infinitely many new sentences." And mental states possess systematicity, because understanding one sentence depends on understanding another sentence. Brian Rabern and Derek Ball, *The science of meaning: Essays on the metatheory of natural language semantics* (Oxford: Oxford University Press, 2018), 29.

[68] He stresses that "computation presuppose a medium of computation: a representational system." Jerry A. Fodor, *The Language of Thought* (Cambridge, MA: Harvard University Press, 1975), 27.

[69] Fodor later proposes in the modular theory of mind that lower-level sensory representation systems are encapsulated as information in higher level systems in a modular form being prepared for higher level cognitive processes such as decision making. Cf. Jerry A. Fodor, *The modularity of mind: An essay on faculty psychology* (Cambridge, MA: MIT Press, 1983)

comprehensible causal sequences by using the syntax of the "language of thought".[70] The language of thought hypothesis presupposes the compositional semantics of mental representations, i.e., the semantics of a complex representation depends on the primitive representations that compose it and its syntactic structure. Semantics in the sense of referential relations can be characterized by the composed complex representations, although they do not necessarily point directly to the external world, but to the causal sequence effects of syntactic structures. That is, Fodor views the meaning of an expression as ultimately exhausted by its referents, and the process of composing referents into causal sequences is the computational process of mental representations. In his words, "mental processes are causal sequences of tokenings of mental representations" and the mind thereby can be regarded as a "syntax-driven machine".[71]

This account of the mind is what Thagard terms the "Computational-Representational Understanding of Mind", with the main assumption that "mind (mental representation + computational procedures = thinking)" is analogous to "program (data structures + algorithms = running programs)".[72] As we have seen, the relationship between the mind and the brain can be seen as analogous to the relationship between a program and a computer, with cognitive processes playing a similar role to the computational processes of hardware running software. Programs can be stored both as data structures in the hardware and as fixed instructions or rules for manipulating other inputs. The main feature of thinking, understood here as the ability to represent external reality by internal symbols through computational procedures, can then be realized by processing structured data with appropriate algorithms. Algorithms are viewed as computable syntactic structures in a language, and their manipulation of symbols is the process of semantic generation.

With the widespread influence of the analogy between minds and computer programs, the debate over whether AI development aims to imitate human thinking or to implement it becomes intense. In Searle article *Minds, Brains, and Programs* (1980), he introduces the terms "weak AI" and "strong AI" to characterize two different assumptions of AI research.[73] The assumption of weak AI is that computers can simulate human intelligence and can be used to help study the mind. In contrast, strong AI is the assumption of computational theory of mind, that computers can have the same mental states and cognitive processes as humans. More specifically, Searle points out that semantics is at the centre of the debate between strong and weak AI. If there is a program that can tell a story and answer relevant questions, those who hold the strong AI hypothesis will argue that the program does understand the meaning of the story, while those who hold the weak AI hypothesis will

---

[70] Fodor, *The Language of Thought*, 36.

[71] Jerry A. Fodor, *Psychosemantics* (Cambridge, MA: MIT Press, 1987), 17, 20.

[72] Thagard, *Mind: Introduction to Cognitive Science*, 11.

[73] John R. Searle, "Minds, Brains, and Programs," *Behavior and Brain Science* 3, no. 3 (1980): 419.

not.[74] To reveal that the strong AI thesis is not convincing, Searle designs the Chinese Room Argument thought experiment.

Suppose there is a person who knows only English and does not speak Chinese, isolated in a room. There is a book of rules written in English in the room, and by following these rules, this person can find the corresponding Chinese characters and the way to combine them. In this way, this person can take Chinese questions from outside the room and get answers in Chinese by using the book of rules, just like a computing process that manipulates the input according to the instructions and gets the output. Eventually, seen just from the outside of the house, the person in the room seems to know how to carry on a conversation in Chinese. This thought experiment attempts to provide a figurative description of how Turing machines are constructed and operate. A Turing machine can theoretically consist of a store that records information, an execution unit that performs instructions, and a control unit that ensures correct execution. When the person uses the rulebook for symbol manipulation, the process of state change in the room is just like how a Turing machine works, performing instructions on the memory through the execution unit and the controller. Searle concludes that the person in the room obviously does not know Chinese and has no idea what the conversation means, so the assumption of strong AI is wrong.[75]

In which sense does the person, or the whole room including the person and the rulebook, not understand the conversation? Searle writes, "whatever purely formal principles you put into the computer, they will not be sufficient for understanding, since a human will be able to follow the formal principles without understanding anything."[76] This argument is obviously targeted at the computational representational understanding of mind where meanings are accounted for as formal syntactic combinations. Further, Searle rejects the multiple realizability of mental states as proposed by functionalism. He denies that materials other than those with biological characteristics may bring about mental states with semantic content, and argues that content is essentially an intentional property of mental states.[77] In his distinction between intrinsic and derived intentionality, intrinsic intentionality is understood to be based on neurophysiological processes in the brain, while derived intentionality is the linguistic competence derived from the intrinsic.[78] Accordingly,

---

[74] Ibid., 420.

[75] Ibid., 418. Searle lists a series of rebuttals, including the system reply that views the entire room as capable of understanding, the robot reply that adds a sensory system to the room, etc.

[76] Ibid.

[77] John R. Searle, "Intentionality and Its Place in Nature," *Dialectica* 38, no. 2/3 (1984): 88.

[78] Ibid., 90. Haugland questions the plausibility of the distinction between intrinsic and derived intentionality. He summarizes Searle's argument as the idea that only physical systems with the correct causal force can have intrinsic intentionality. He uses a thought experiment to demonstrate that Searle's view is in fact a property dualism. Imagine a demon who takes all the neurotransmitters that Searle attributes to the human biochemical structure, and this demon is placed in the human brain and covers the neurons with the correct causal force. Yet according to Searle, the demon still does not have any intrinsic intentionality. This shows that Searle either admits that having the right causal powers is not sufficient to have intrinsic intentionality, which would lead to

only organisms with a neurobiological basis can possess mental states, and thus semantic contents. Semantics can therefore only be attributed to systems that have identical causal processes as the brain, i.e. that have the right causal power. To the question "Can machines think?", Searle answers yes and indicates that humans are precisely such thinking machines, but a kind of biological machine that works differently from computers.[79] This shows a biological naturalistic view that minds can only be caused by brains.[80]

This view recalls Ryle's criticism of the category mistakes committed by those who, on the one hand, distinguish between mind and machine as two categorical things and, on the other hand, try to explain mind by seeking a new causal hypothesis in terms analogous to mechanical laws.[81] Indeed, functionalism is arguing against the substantializing treatment of the mind in the identity theory. Unlike identifying mental phenomena with neurophysiological phenomena, functionalism models mental states as mechanism of information processing. The validity of this modeling is demonstrated precisely by realizing what mental phenomena can do with other physical systems. Although Searle demonstrates that the functionalist approach is incapable of realising an identical mind to the one caused by the human brain, the strong AI thesis that achieves the same causal effects as the human brain actually misunderstands the functionalist approach. In the computational representational version, the mind is viewed as a computational system, which is not equivalent to a physical machine, but rather a model of information processing. The misconception that a human brain is equivalent to an electronic computer is due to misplacing the theorization or modeling of cognitive processes into physical space, either by looking for a correspondence between mental and physical states, or by looking for the physical systems in which mental processes reside. These attempts are akin to looking for the "ghost in the machine".

The functionalist approach attempts to provide an explanation of how the mind works without having to depend on the causal relations between physical and psychic states. The debate over the strong AI thesis is an extended issue of replacing the mind with how the mind works as the research object. A more practical challenge for functionalism is the limitations of symbolic AI based on based on well-defined data structures and fixed rules of symbol manipulation. Although symbolic AI approaches have developed from an early focus on explicit representations of

---

an unbridgeable explanatory gap between causal powers and intentionality, or a property dualism. Or, if Searle unifies that correct causal powers can lead to mental phenomena, then he holds the same brain-as-machine assumption as the computational theory of mind. John Haugland, "Programs, causal powers, and intentionality," *The Behavioral and Brain Science* 3 (1980): 432.

[79] Searle, "Minds, Brains, and Programs," 422.

[80] Cf. John R. Searle, "Biological Naturalism," in The Blackwell Companion to Consciousness, eds. Max Velmans and Susan Schneider (Oxford: Blackwell, 2007), 327–336.

[81] John R. Searle, "Why I am not a property dualist," *Journal of Consciousness Studies* 9, no. 12 (2002): 61. He later changed his strategy to defend consciousness as a first-person subjective experience from a social ontological perspective, in distinction to a third-person neurobiological process.

well-defined tasks and solutions to representations of domain-specific knowledge and its reasoning, they still fail to be applied to domains with incomplete knowledge and flexible task environments.[82] This shows that if symbolic manipulation is performed simply according to given rules, then information processing systems cannot learn how their internal states relate to the external world. Fixed coding of functions misconstrues the dispositional powers of the mind. Against this background, connectionism is revived and sub-symbolic AI approaches rise.

2.2.3 Sub-Symbolic AI, Connectionism, and the Unexplainable Issue

Sub-symbolic AI can be traced to the tradition of connectionism, the neural network-based approach to modeling cognitive processes.[83] The connectionist version of the computational theory of mind holds that the key to modeling cognition is how neural networks are organized into computational models without giving fixed computational procedures in advance. On the one hand, this requires plasticity in the information processing architecture, in that it is trainable and capable of "learning". On the other hand, pattern recognition, i.e., the processing sensory inputs such as images, characters, and audio, was regarded as based on the co-ordination of perceptual, memory, and other experiences, rather than on innate rules.[84] Thus, early applications of neural networks were to model perceptual processes, until they were reintroduced in the 1980s when symbolic AI failed to make expected progress in areas such as machine translation.[85]

An artificial neural network is an architecture that simulates the effects of synapses between neurons. This model consists of neuron-like nodes and synaptic weights presenting the strength of nodes' connections. The weights of the node

---

[82] Studer Rudi, Benjamins V. Richard and Dieter Fensel, "Knowledge engineering: Principles and methods," *Data & Knowledge Engineering* 25, Issues 1–2 (1998): 162.

[83] Connectionist approaches were prevalent before the dominance of symbolic AI. The logical model of neural networks proposed by McCulloch and Pitts in 1943 can be seen as the origin of both classical computational theory of mind and connectionism. In the early 1950s, Minsky built the SNARC system, the first neural network simulator. These approaches were inspired by neurophysiological studies. Stuart J. Russell and Peter Norvig, *Artificial Intelligence: A Modern Approach* (Harlow: Pearson Education Limited, 2016), 17.

[84] Rosenblatt's work on perceptron is one of the earliest artificial neural network architectures that models neural networks for perceiving, learning, and memory. Cf. Nilsson, *The Quest for Artificial Intelligence,* 92 ff.

[85] Connectionist and symbolic AI based on logical reasoning approaches coexisted for some time as different research directions. Yet by the mid-1960s, symbolic AI gradually dominated the research field. One reason for this is the limited ability of neural networks built on a single perceptron as an expected model for cognition. In 1969 Minsky and Papert published *Perceptrons: An Introduction to Computational Geometry,* stating the limitations of the model of perceptron. Ibid., 262. Cf. Lieto, *Cognitive Design for Artificial Mind*, 14f.

connections are variable during training. Neural networks can be architected like a cascade of perceptrons that process information in a parallel distributed manner, including one or more hidden layers in addition to input and output layers. Correspondingly, multilayer neural network-based machine learning is known as deep learning, which "allows computational models that are composed of multiple processing layers to learn representations of data with multiple levels of abstraction".[86] In a multilayer neural network, each layer can represent data features in a particular structure, which in turn can be learned and represented at more abstract levels.

In such trainable information processing models, neuron-like nodes are "sub-symbolic" without fixed structure, while neural networks are learned "emergent systems".[87] Unlike symbolic AI, which processes information top-down according to program instructions, the neural network model processes information bottom-up, distributing information over different connection weights. To train the neural network efficiently, a learning algorithm called "backpropagation" was introduced.[88] This algorithm is a form of supervised machine learning, whereby a neural network is trained with multiple input sample data and desired outputs in order to generalize a model that can handle new inputs.[89] That is, instead of relying on explicit program instructions, information processing relies on generalizing and optimizing models from the data while training. The success of this statistical-based method challenges rule-based symbolic AI approaches. While symbolic AI approaches use symbols to fix syntax, sub-symbolic approaches treat neuron-like nodes as tools that can be connected into different symbolic structures. This plasticity allows sub-symbolic AI approaches to be more effective in many application areas.

With respect to cognitive science and the concept of intelligence, connectionism argues against the separation of "lower-order" sensory systems from "higher-order" cognitive processes. It reveals the irreconcilability between hypotheses of internal cognitive structures such as Fodor's language of thought and perceptual plasticity. Fodor considers sensory input systems as modules that informationally encapsulate basic representation for higher cognitive processes.[90] Sensory input is thought to be

---

[86] Deep learning is a form of feature learning, also called representation learning, allowing machines to learn the features of raw data and transform them into available representations. Yann LeCun, Yoshua Bengio, and Geoffrey Hinton, "Deep learning," *nature* 521, no. 7553 (2015): 436.

[87] Lieto, *Cognitive Design for Artificial Mind*, 13ff.

[88] Ibid., 15.

[89] With this method, McClelland and Rumelhart trained a neural network that can predict the regular or irregular past tense of English verbs. Cf. Rumelhart, David E. and James L. McClelland, 1986, "On Learning the Past Tenses of English Verbs", in *Parallel distributed processing: explorations in the microstructure of cognition, vol. 2: psychological and psychological and biological models* (Cambridge, MA: MIT Press, 1986), 216–271.

[90] "Informationally encapsulation" means that "the data can bear on the confirmation of perceptual hypothesis includes [...] the confirmation function for input systems does not have access to all of the information that the organism internally represents; there are restrictions upon the allocation of internally represented information to input processes." Fodor, *The modularity of mind: An essay on faculty psychology*, 69.

acquired neutrally and then used for theorizing. Against this, Churchland demonstrates that cognitive processes are not impermeable to perception based on research evidence from neuroscience.[91] The plasticity of perception shows that there is no neutral observational knowledge. This view holds that cognitive processes do not follow inherent rules and that the environment plays an active role in shaping cognitive processes. Intelligence is understood here as the ability to adapt to the environment and learn.

Moreover, the modeling of perception in terms of statistical and probabilistic approaches has gradually become mainstream in cognitive science. In fact, the view of perception as a probabilistic reasoning process can be traced back at least to Helmholtz. This view was reintroduced in computational neuroscience as the Bayesian brain hypothesis, i.e., the hypothesis that the function of the brain is to predict information about the external world in response to its uncertainty, and that perception is testing predictions.[92] The predictive process involves using Bayes' theorem to infer sensory information and in turn combining current sensory input with a priori expectations to update predictions. Because this is a continuous induction of predictive models along with deductive reasoning, it is also known as "analysis by synthesis".[93] This process is thought to match the plastic connectivity of neural networks.[94] As one model for the implementation of the Bayesian brain hypothesis, "predictive coding" have become a dominant paradigm in cognitive science over the last decade, based on modeling perception as an "error correction" plastic process.[95] Predictive coding is originally a technique of compressing image data. Since the values of neighboring pixels of an image can usually be predicted, predictive coding strategies encode only unexpected prediction errors, such as graph boundaries. This allows efficient compression or transmission of images without encoding all pixels, processing only the divergence from the predicted signal. Using such models in neuroscience means making "prediction error into a kind of proxy for sensory information itself".[96] Predictive coding models are considered to be closer to perceptual systems than the back-propagation method, because predictive coding does not require additional training samples and a supervised training process. Rather, the prediction when receiving sensory signals itself is the training. Predicting signals by theorizing prior experience can be well integrated with error

[91] Paul M. Churchland, "Perceptual Plasticity and Theoretical Neutrality: A Reply to Jerry Fodor," *Philosophy of Science* 55, no. 2 (1988): 169 ff.

[92] Karl J. Friston, "The history of the future of the Bayesian brain," *Neuroimage* 62, no. 2 (2012): 1231.

[93] Andy Clark, "Whatever next? Predictive brains, situated agents, and the future of cognitive science," *Behavioral and Brain Sciences* 36, no. 3 (2013): 182.

[94] Friston, "The history of the future of the Bayesian brain," 1231.

[95] Clark, "Whatever next? Predictive brains, situated agents, and the future of cognitive science," 181.

[96] Ibid., 183.

correction methods, allowing perception to be modeled as a bidirectional cascade that predicts the current sensory input and corrects prediction errors.[97]

Computer simulation of cognitive processes relies significantly on the architecture of artificial neural networks and their learning mechanisms. In terms of architecture, multilayer neural networks such as convolutional neural networks have shown the advantages of training large amounts of data over the last decades. Compared to early neural network research that had only a few intermediate hidden layers, deep neural network architectures can reach thousands of hidden layers. In terms of training mechanisms, traditional machine learning methods can be divided into supervised and unsupervised learning according to the way of training data. The former requires sample data for training, while the latter does not need labeled data but rather discovers latent relationships among them during training. There are also other training methods such as reinforcement learning, training a decision agent to act through trial and error in an environment, receiving rewards or punishments to learn optimal action strategies.[98] Incorporating reinforcement learning into the training of neural networks has led to significant advances in AI technologies.[99]

Since the connectionist revival, the expectations and concerns about general-purpose AI technologies have risen along with the emergent theory. According to the emergent theory, human intelligence can be viewed as an emergence from the structure of the brain, just as the biological functions of organic molecules can be seen as an emergence of the chemical structure of the molecule.[100] Emergence is not the integration of parts into a whole system, and emergent properties or functions are not reducible to the original system mechanism. This irreducibility leads to

[97] Ibid., 187. "Perception here becomes 'theory-laden' in at least one (rather specific) sense: What we perceive depends heavily upon the set of priors (including any relevant hyper priors) that the brain brings to bear in its best attempt to predict the current sensory signal." (187)

[98] Ethem Alpaydin, *Introduction to Machine Learning* (Cambridge, MA: MIT Press, 2014), 517.

[99] AlphaZero is one impressive example of combining deep convolutional neural networks with reinforcement learning. AlphaZero is a game program developed by DeepMind that learns and play chess, shogi and go. Unlike previous programs that need to learn on the basis of board game records or other experiences from human players, AlphaZero learns by playing against itself. The winning side in the self-play game is trained as the best player for the next game. After a short period of training, it can reach the level of AlphaGo, which has defeated the top human Go players. Since AlphaZero has mastered three game types with a set of general-purpose learning algorithms rather than being restricted to domain-specific knowledge, such remarkable results raise expectations for general-purpose learning algorithms. However, the complex structure of such models makes it difficult to explain some of the ways it learns to make decisions that are beyond human intuitive thinking. David Silver et al., "A general reinforcement learning algorithm that masters chess, shogi, and Go through self-play," *Science* 362 (2018): 1141ff.

[100] The emergent theory is divided into the "strong emergent" hypothesis and the "weak emergent" hypothesis. Strong emergence holds a biological naturalistic position, while weak emergence theory is similar to computational functionalist account of intelligence. More specifically, strong emergence refers to intelligence as a result of inherent properties of interacting biological neurons and brains, while weak emergence refers to intelligent functions that can be simulated by computer programs and computer circuit models. Joscha Bach, "Seven Principles of Synthetic Intelligence," in *Artificial General Intelligence*, ed. Pei Wang et al. (Amsterdam: IOS Press, 2008), 69.

difficulties in bottom-up engineering. One can neither know exactly when nor how intelligence will occur. In such a speculative way, emergent theory raises expectations for the emergence of human-like general AI on the one hand, and on the other hand worries about the emergence of superintelligence that no longer adheres to human norms.

Artificial general intelligence (AGI) is used to outline the assumptions of general AI, which aims to develop AI systems that are capable of learning to perform any task performed by humans. The basic assumption of AGI is that the modeling of cognition can be unified so that it is possible to implement various aspects of intelligence within one framework, incorporating the "narrow AI" that is usually developed by specialized domains.[101] Yet the challenges encountered by AGI are not only method-specific, but conceptual. Although its goal seems clear, namely, to achieve human-level general intelligence, it is difficult to find a unified metric and roadmap for reaching this goal.[102] To measure whether the capabilities of AGI agents are sufficiently general, one view is that it is necessary to find suitable tasks and "scenarios" that can test various aspects of intelligence. DeepMind's development of XLand is an implementation of this view.[103] XLand simulates a universe of scenarios containing various task environments, in which AI agents develop skills through reinforcement learning to accomplish different tasks. The goal of this research is to develop AI agents with general capabilities in the simulated universe of scenarios.[104] By keeping the training tasks open and dynamic, the AI agents can acquire some general skills in successive iterations, such as simple use of tools in trial and error and cooperation with other agents. This approach is thus also called open-ended learning, allowing artificial agents to keep learning in simulated task environments.

Another possible pathway to AGI is the currently highly successful and controversial large language models, represented by OpenAI's GPT family, DeepMind's Gato, Google's PaLM, etc. Unlike the above method of training general-purpose artificial agents in virtual task scenarios, large language models, as a generative AI model, generate outputs with similar characteristics of human-generated data by

[101] In the past AGI research, there have been various attempts to build general intelligent systems, including the CYC project to represent common sense knowledge with first-order predicate logic, the ACT-R framework to simulate human psychology, the NARS system to build uncertainty-based symbolic AI systems with fuzzy logic, using Bayesian networks to deal with general uncertainty problems, etc. Although numerous narrow AI technologies are already being used in many areas of industry and common life, few researchers in narrow AI consider their work as oriented toward AGI because of the difficulty of integrating all the subfields together. Cassio Pennachin and Ben Goertzel, "Contemporary Approaches to Artificial General Intelligence," in *Artificial General Intelligence* (Berlin, Heidelberg: Springer Berlin Heidelberg, 2007), 1–6.

[102] Sam Adams et al. "Mapping the Landscape of Human-Level Artificial General Intelligence," *AI Magazine* 33, no. 1 (2012): 29.

[103] Open Ended Learning Team, Adam Stooke et al. "Open-ended learning leads to generally capable agents," *arXiv preprint arXiv:2107.12808* (2021).

[104] Ibid.

learning the patterns and structures of the training data.[105] The ability to generate human-like output is highly dependent on its training datasets and post-training tuning. In other words, the ability to generate human-like outcomes does not emerge unconditionally. I will specify in Chap. 6, with the GPT family as a case study, why the large language model approach is a feasible path to AGI, and what it still lacks. I will also argue that the expressive power to engage in discursive practices enables intelligent agents to recognize each other. These refute the idea that the formation of intelligent agents can be explained by the emergence of intelligent systems on their own.

In fact, the emergent theory not only fail to adequately address how human-level AI could emerge, but also lead to myths about intelligence being divorced from human understanding and control. The assumption of superintelligence is summarized by Bostrom as the possibility of building machine brains based on the human brain.[106] He details several paths to superintelligence, such as AI path, whole brain emulation, biological cognition, brain-computer interface, networks, and organizations.[107] According to his assessment, superintelligence may be achieved first by means of AI, especially AI technologies based on simulating the human brain can gradually achieve the simulation of the entire nervous system starting from parts.[108] Once an artificial physical system surpasses the human brain in capability, there may emerge a superintelligence capable of escaping human control and threatening humanity. This superintelligence hypothesis evokes fear of AI technologies, especially as sub-symbolic AI rapidly develops with inexplicable algorithms.[109] In recent years, the concept of superintelligence has been combined with the notion of "technological singularity", which refers to a point when technological change accelerates to an explosion, and along with it, there will be an explosive cumulative growth of intelligence. By then, it is expected that artificial intelligence will also exceed the level of human intelligence.

But how is it possible for artificial intelligence, which originates from the imitation of human intelligence, to transcend human intelligence? Chalmers argues that three premises will lead to the conclusion that such superintelligence will emerge: (1) the "equivalence premise" that there will be fully human-level AI created by human intelligence; (2) the "extension premise" that there will be AI+ created by

---

[105] Cf. Mark Coeckelbergh and David J. Gunkel, ChatGPT: deconstructing the debate and moving it forward," *AI & Society* (2023).

[106] Nick Bostrom, *Superintelligence: paths, dangers, strategies* (Oxford: Oxford University Press, 2014), vii.

[107] Ibid., 22–51.

[108] Ibid., 50.

[109] Indeed, the hypothesis of super intelligent machines was already proposed in the early days of AI research. The thought was that if there is an intelligent machine with engineering capabilities, then it can improve itself or create more intelligent machines. One of the earliest versions of the super intelligent machine hypothesis can be traced back to Irving John Good's "Ultraintelligent Machine" speculation in 1965. See David J. Chalmers, "The Singularity: A Philosophical Analysis," *Journal of Consciousness Studies* 17 (2010): 7.

human-level AI, and better than it; and (3) the "amplification premise" that there will be AI++ created by AI+.[110] One of the main arguments for the first premise is the possibility of human brain emulation, i.e., the human brain is a machine, and if we emulate such a machine, there will be human-level AI. It assumes that intelligence emerges from physical systems and can be extended or augmented depending on the accumulation of physical systems. As a result, a superintelligence that far exceeds human intelligence is conceivable.

In this context, intelligence is viewed as an irreducible but cumulative capability that emerges as physical systems increase in size and complexity, just as computing power is to computers. This connectionist account of emergent theory analogizes plastic brains to adaptive artificial neural networks. An intelligent system does not perform tasks according to given rules, but rather has learning capability and acquires appropriate functions during training. However, this learning capability based on the computational power of physical systems does not explain their learning processes by itself. If learning processes refer only to the emergence of intelligent systems from physical systems, then such intelligent systems would become black boxes that can only be identified by input and output relationships. Put differently, explaining learning capability only in terms of the computational power of a physical system does not explain how an intelligent system acquires appropriate functions, because the emergent intelligent systems surpass the explanatory framework of original systems. The problem of the spontaneous emergence of unexplainable intelligent systems raises warnings about AI safety and ethical issues.

As a response, explainable AI (XAI), also called explainable machine learning, has become a research orientation dedicated to developing interfaces that can explain how machine learning processes make decisions for end users, particularly in normative areas involving legal and ethical issues.[111] The goal of explainable AI is to enable end-users to understand the rationale for decisions made by AI systems and to appropriately trust and effectively manage AI systems.[112] This requires AI systems to learn users' requirements for effective explanations in their interactions with humans. This orientation signals a normative turn in AI development, as explainable AI is concerned with how users can trust AI and who is responsible for decision-making, which cannot be seen as purely technological issues independent of social norms.

The normative turn is necessary and urgent for the development and application of AI technologies. Without the guidance of a normative framework, increased computing power only triggers fear and mistrust. Unlike turning a black box into transparent, explainable technology involves not seeing the inside, but rather the

[110] Cf. Ibid., 14ff.

[111] DARPA, "Broad Agency announcement: explainable artificial intelligence (XAI)," *DARPA-BAA-16-53*. (Defense Advanced Research Projects Agency, 2016), 6ff.

[112] David Gunning and David Aha, "DARPA's Explainable Artificial Intelligence (XAI) Program," *AI Magazine* 40, no. 2 (2019): 45.

accountability across the technological system.[113] Floridi points out that an urgent principle for AI technologies is explicability, which can be expressed in two ways: *epistemically intelligible* and *ethically accountable*.[114] The explicability principle underlines that the development and application of AI technologies requires more than just explaining their complex functioning, but also translating them into problems that can be dealt with in everyday practice. This means explaining is not only a technical question of how to account for the black box, but integrating expertise into the framework of social life. This normative turn, then, requires a rationalist approach to demystify intelligence in terms of computational power only.

2.3 Rethinking the Relation Between Minds and Machines in Logical Space

So far, we have seen a series of analogies in AI development and cognitive research, from analogies between organic and circuit systems, between minds and computing models, to between human brains and machine brains. In cybernetics, a self-regulating circular feedback system is seen as a universal model based on information communication, describing biological self-organization and electronic self-control systems. In the cognitivist turn, the information processing of human cognition is regarded as a more complex symbolic manipulation process. Symbolic manipulation is not only about the reception and feedback of information, but also about the way cognitive systems process information according to inherent syntax-like structures. With this background, symbolic AI approaches develop programs that perform tasks according to fixed rules, mimicking how humans process symbols in a particular way. However, symbolic AI approaches are limited in their application since they rely on well-defined data structures and given rules. As a response, sub-symbolic AI approaches train artificial neural networks to learn and optimize the procedure for processing information. Information processing rules are acquired during the learning process rather than given in advance. The learning process of artificial neural networks models the plasticity and the connectivity of the brain. But in this approach, an intelligent system is treated as a black box that is identified in training, and how it emerges is not explained by the original system. It becomes a

[113] Ananny and Crawford argue that behind the idea of "making the black box transparent" there is an assumption that what can be seen should be held accountable. This idea oversimplifies the problems that arise in decision-making by AI systems, such as bias, by transferring them to problems that can be solved by technology alone. Mike Ananny and Kate Crawford, "Seeing without knowing: Limitations of the transparent ideal and its application to algorithmic accountability," *New media & Society* 20 (2018): 973–989. I'll return to the bias problem and the discussion of why technology can't solve it on its own in Sect. 6.2, using ChatGPT as an example.

[114] Luciano Floridi and Josh Cowls, "A Unified Framework of Five Principles for AI in Society," *Harvard Data Science Review* 1, no.1 (2019): 8.

challenge to combine the advantages of symbolic AI and sub-symbolic AI approaches.

Behind these analogies, there is a general assumption from the functionalistic computational theory of mind, that is, the mind can be simulated through computational models, whether Turing machine style, connectionist, or other information processing theories. This assumption serves as a core methodology motivating AI research and cognitive science. Human intelligence and artificial intelligence are analogous to each other, and they both serve as the object being investigated and as the reference for research and development. If the analogies are not taken as a normative guide to research, but are treated as mutual evidence in a naturalistic way, it leads to begging the question in elaborating the concept of intelligence.[115] Modeling of information processing would correspondingly be misunderstood as a reductionist method, such as reducing the conditions for configuring and performing proper information processing procedures to the computational power of the brain or hardware.

The analogical method is itself a manifestation of human intelligence. As explained in Sect. 2.1, the performance of intelligent capacities is an exercise process, and the mind can be understood as a series of dispositional propositions that can be tested and adapted as semi-hypotheses. Yet the semi-hypothesis that "as if" is true in modeling is often misunderstood as a substantive premise. If the analogical approach is misinterpreted as an actual categorization, it leads to category mistakes. As Ryle clarifies the misunderstanding of mind, the category mistake begins with treating the mind as a different "thing" from matter, while believing that the two different "things" are categorically equivalent and possess the same type of properties. The inner mind and the public body can only occur as two things in two different worlds. According to this diagnosis, treating the mind as an entity like the body turns the inquiry into the place of the mind in the physical world into the search for the ghost in the machine. With this methodological misdirection, one explores the occurrence of the mind within a physical framework on the one hand and finds this to be an impossible task on the other. The myths about artificial intelligence inherit the myths about human intelligence.

First, the computational theory of mind, in either its representationalist or connectionist version, may be taken as the myth that once a physical system achieves sufficient computational power, it may become a super intelligence that can do anything on its own. The myth of equating intelligence with the computational power of a system ignores the fact that aspects of human intelligence are realized with the

[115] Shuman and Du explain in detail in their paper that using each other as references for one's own research object may lead to a circular argument. On the one hand, the concept of AI is referenced to intelligence that has been reduced to what can be simulated by computers, and on the other hand, the cognitive science of the mind is referenced to such an approach to AI. This mutually held reference may lead to a "petitio principii". Namely, the two propositions that "the mind is a computer" and that "a computer can simulate the mind" are made to prove each other. Nadine Schumann and Yaoli Du, "Grenzgänge: Von Menschen zu smarten Maschinen – und zurück?" *Künstliche Intelligenz – Die große Verheißung*, eds. Anna Strasser et al. (Berlin: Xenomoi Verlag, 2021), 52.

ongoing experience of practice in its environment. This myth presupposes a substantive physical system and computational power as its inherent property. Accordingly, explaining how a computational model works is not about revealing causal effects within an entity, but about the conditions under which an intelligent agent is trained as such. The training method and environment become significant factors in explaining intelligent agents.

Second, the myth of intelligence arises from the opposition between the natural and the artificial, between naturalism and normativism. As we have seen, the main argument of biological naturalists is that only an exact duplicate of the brain can have intrinsic intentionality and is thus genuinely intelligent.[116] This naturalist view thus considers intelligence to be a property or function that emerges when a system achieves a certain level of performance capable of yielding specific causal effects. This leads to the idea that there could be intelligent agents that are exact replicas of the human brain yet may not understand human values because it is not shaped by human social norms. However, as Turing repeatedly points out, the process by which machines are trained to go from unorganized to organized universal machine models is a process of developing discipline, much like the process of educating a human being. And the education of a human being in everyday life is not just about refining intelligence, but also acquiring ways of acting and thinking that meet social norms. The analogy between learning machines and educating persons emphasises the social and normative nature of human intelligence.

AI, as an engineering goal, must become capable of being explained within a normative framework in order to be applied by human users and thus realize their intelligent behaviors. Taking this into account, the goal of AI development should be specified as agents of rational action. To clarify what kind of artificial agents are intelligent, Russell and Norvig offer four candidates: *thinking humanly*, *acting humanly*, *thinking rationally*, or *acting rationally*.[117] The first two approaches focus on "human-like", while the latter two focus on rationality. As we have seen in this chapter, myths about human intelligence can mislead researches aiming at "human-like" artificial agents. The third candidate, the rationalist approach to thinking, is the search for "laws of thought" assumed to be prerequisites for intelligent behaviour. But this approach does not account for how the laws are formed, as in the case of symbolic AI operating according to given rules. The final candidate, the rationally acting agent, is then proposed by them as the goal of developing AI, where rationality is defined as the limited rational actions that an agent can achieve in its environment.[118] In this development orientation, intelligent action is not the embodiment of some universal law of rationality, but the appropriate ways of processing information in specific environments. Thus, the concept of intelligence can be outlined as an

[116] I will introduce in Sect. 3.3.1 Verbeek's conception of material intentionality and in Sect. 5.2 Gallagher and Miyahara's proposal for operative intentionality, which is not centered on internal brain states but distributed across the interaction between the agent and the environment.

[117] Russell and Norvig, *Artificial Intelligence: A Modern Approach*, 2.

[118] Ibid., 5.

agent's ability to achieve goals in a wide range of environments.[119] Intelligence understood in this agent-environment framework is situated. Intelligent agents learn how to reach goals and carry out norms in their environments. In this sense, intelligence is no longer an emergent property of a system, but rather the disposition of the situated agent to realize rational action.

In the computational representational understanding of mind, thinking (mental representations + computational procedures) is analogous to running programs (data structures + algorithms), where mental representations are assumed to be explicitly structured according to language-like syntax, and they are stored as fixed data structures.[120] But mental representations modeled by well-defined stored data structures lack the flexibility to follow rules and update in response to environments. Indeed, Ryle's diagnosis of intellectualism has shown that procedural knowledge arising in practice underlies propositional knowledge, not vice versa. Mental states are accordingly dispositions to act intelligently, without having to be understood as internal representations. In such a disposationalist view, mental processes can be expressed in terms of logical linguistic structures, without the need to assume internal representational structures. That is, mental processes do not need to be modeled as well-defined data structures and given algorithms. Rather, structuring data is learning and enacting procedures for information processing in the environment.

Yet if intelligence becomes a dynamic, relational process between agents and the environment, how can the Rylean approach explain the internal thinking episodes that are usually considered to be intelligent? One answer to this can be found in an "anthropological science fiction" as thought experiment conceived by Sellars. In Sellar's thought experiment, there are some "Rylean ancestors" who use a behavioristic language that can express and explain public objects and events, but have no available linguistic expressions about inner episodes.[121] So how is it possible for the Rylean ancestors to recognize that others and themselves are thinking? Sellars proposes that expressing inner episodes can be understood as making theoretical postulations about them by means of linguistic structures.[122]

[119] Shane Legg and Marcus Hutter, "Universal Intelligence: A Definition of Machine Intelligence," *Minds & Machines* 17 (2007): 402.

[120] The representationalists' conceptualization of "storage" can be compared to the concept of "long-term memory store" in memory theories. Eric Schwitzgebel, "The Nature of Belief from a Philosophical Perspective, With Theoretical and Methodological Implications for Psychology and Cognitive Science," *Frontiers in psychology* 13: 947664 (26 Jul. 2022), 1.

[121] Wilfrid Sellars, "Empiricism and the Philosophy of Mind," 309.

[122] Sellars raises the question of what needs to be added to the Rylean language so that these talking animals can recognize each other as thinking animals. He begins by suggesting that adding some semantic vocabulary allows people to express true and false in their discourse. In this way, expressions such as "mean" and "imply" are distinguished from causal expressions of behaviorism. This seems to allow one to express "thoughts" in a common sense, since thoughts are here isomorphic to indicative expressions, both refer to something. He notes: "For characteristic of thoughts is their *intentionality*, reference, or aboutness, and it is clear that semantical talk about the meaning or reference of verbal expressions has the same structure as mentalistic discourse concerning what

Suppose an ancestor of the behaviorist linguistic community observes that one can behavior intelligently without expressing it in words, he might invent a theory that overt speech begins with inner speech. Namely, once a language user becomes skilled in the way that language is used internally, inner speech can be used to express inner episodes in a way that appears to be a direct report of them. This is, however, a typical behaviorist misunderstanding of language that considers speech to be a report of events, and speech of inner episodes becomes a privileged report.[123] Sellars points out that "the reporting role of these concepts—the fact that each of us has a privileged access to his thoughts—constitutes a dimension of the use of these concepts which is *built on* and *presupposes* this intersubjective status".[124] The distinction between inner and public events is not a report based on observation, but a theoretical categorisation. The privileged report of inner episodes is based on the presumption that everyone has privileged access to their own thoughts. And this presumption requires recognising both oneself and others as persons with this privilege.

If Sellars is right that reports are in fact theories endowed with a reporting role, then the explanation of reports about inner episodes should be traced back to the public use of language. Accordingly, thoughts, at least as they can be structured and articulated in language, arise from common discursive practices. Thoughts can be treated as private inner episodes precisely because the intersubjectivity of language use allows for a theoretical distinction between internal and external, between privileged private spaces and naturalized public spaces. Sellars writes:

> As I see it, this story helps us understand that concepts pertaining to such inner episodes as thoughts are primarily and essentially *intersubjective*, as intersubjective as the concept of a positron [...] My myth has shown that the fact that language is essentially an *intersubjective* achievement, and is learned in intersubjective contexts [...] is compatible with the "privacy" of "inner episodes". It also makes clear that this privacy is not an "absolute privacy". For if it recognizes that these concepts have a reporting use in which one is not drawing inferences from behavioral evidence, it nevertheless insists that the fact that overt behavior *is* evidence for these episodes *is built into the very logic of these concepts*, just as the fact that the observable behavior of gases is evidence for molecular episodes is built into the very logic of molecule talk.[125]

---

thoughts are about. It is therefore all the more tempting to suppose that the *intentionality* of thoughts can be traced to the application of semantical categories to overt verbal performances, and to suggest a modified Rylean account according to which talk about so-called '*thoughts*' is shorthand for hypothetical and mongrel categorical-hypothetical statements about overt verbal and nonverbal behavior, and that talk about the intentionality of these 'episodes' is correspondingly reducible to semantical talk about the verbal components." Ibid., 180.

[123] Ibid. 189.

[124] Ibid.

[125] Ibid. It also shows that science as a human enterprise cannot be divorced from common practices. Sellars writes: "The truth of the matter [...] is that science is continuous with common sense, and the ways in which the scientist seeks to explain empirical phenomena are refinements of the ways in which plain men, however crudely and schematically, have attempted to understand their environment and their fellow men since the dawn of intelligence." Ibid., 182.

Linguistic expressions, even if some appear to be direct reports of internal or external events, are framed by the linguistic structure of intersubjective discursive practices, and not merely by individual responses to stimuli. The meaning of an expression thus cannot be reduced simply to the referential relationship between expression and object, but involves a broader framework of intersubjective discursive practices.[126] Talking about internal episodes is thus a form of theorizing in common discursive practices, as Sellars puts it, "these episodes are 'in' language using animals as molecular impacts are 'in' gases, not as 'ghosts' are in 'machines''".[127]

This claim allows for a new understanding of the relation between minds and machines from the standpoint of rational beings who use language. It is a mistake to think of material reality and inner life as two worlds. Human intelligence is not simply representing a given reality but rather representing a meaningful world and realizing itself in the process. Here semantics is not grounded in a correspondent relation between the internal mind and the external world, but is generated in the construction of a logical space that involves material inferences and tacit endorsements of them. This logical space can be articulated and is called by Sellars the "logical space of reasons".[128] In shared discursive practices, true propositions and thoughts compose the logical space for minds that is distinct from the psychological and physical realms, as Frege argues for the objective realm. The normative and logical space structured in discursive practices expresses the mental and material worlds through vocabulary such as inner and outer. Therefore, minds, machines, and their relations should be examined in this space where various possible combinations of dispositional, semi-hypothetical propositions that can be presented and tested. Rational agents and a world shaped by their normative actions are co-constituted in such practices.

If human beings structure such a logical space of reasons in discursive practices and mutually authorize each other to enter it as rational agents, can other information processing agents access such a space? The next chapter will elaborate on how today's technological environment materializes a normativ and logical space of human discursive practices. This new technological environment, supported by the infrastructural architecture of the internet and semantic technologies, can provide information processing agents, whether human or machine, with readable resources and possibilities for rational action.

---

[126] Ibid., 195. Sellars concludes, "though it rests on a framework of discourse about public objects in Space and Time, has an autonomous logical structure, and contains an *explanation of*, not just a *code for*, such facts as that *there looks to me to be a red and triangular physical object over there*."

[127] Ibid., 187.

[128] Ibid., 272. Sellars writes "when we use the word 'see' as in 'S sees that the tree is green' we are not only ascribing a claim to the experience but endorsing it. It is this endorsement which Ryle has in mind when he refers to seeing that something is thus and so as an achievement, and to 'sees' as an achievement word. I prefer to call it a 'so it is' or 'just so' word, for the root idea is that of truth." (272) Based on this account, operations on representations and access to resources can be incorporated into an inferentialist framework, which is compatible with situated cognition. I will discuss material inference and inferentialism in detail in Chap. 5.

Bibliography

Adams S, Arel I, Bach J, Coop R, Furlan R, Ben Goertzel J, Hall S, Samsonovich A, Scheutz M, Schlesinger M, Shapiro SC, Sowa J (2012) Mapping the landscape of human-level artificial general intelligence. AI Mag 33(1):25–42. https://doi.org/10.1609/aimag.v33i1.2322

Alpaydin E (2014) Introduction to machine learning. MIT Press, Cambridge

American Society for Cybernetics (n.d.) Summary: the Macy conference. https://asc-cybernetics.org/foundations/history/MacySummary.htm. Accessed 4 May 2021

Ananny M, Crawford K (2018) Seeing without knowing: limitations of the transparent ideal and its application to algorithmic accountability. New Media Soc 20:973–989. https://doi.org/10.1177/1461444816676645

Ashby WR (1957) An introduction to cybernetics. Chapman & Hall, London

Bach J (2008) Seven principle of synthetic intelligence. In: Pei W et al (eds) Artificial general intelligence. ISO Press, Amsterdam, pp 63–74

Bostrom N (2014) Superintelligence: paths, dangers, strategies. Oxford University Press, Oxford

Bunge M (1963) A general black box theory. Philos Sci 30(4):346–358

Chalmers D (2010) The singularity: a philosophical analysis. J Conscious Stud 17:7–65

Chomsky N (1956) Three models for the description of language. IRE Trans Inf Theory 2:113–124

Chomsky N (1977) Empiricism and rationalism. Excerpted from language and responsibility. Pantheon, New York

Churchland PM (1988) Perceptual plasticity and theoretical neutrality: a reply to Jerry Fodor. Philos Sci 55(2):167–187

Clark A (2013) Whatever next? Predictive brains, situated agents, and the future of cognitive science. Behav Brain Sci 36(3):181–204. https://doi.org/10.1017/S0140525X12000477

Coeckelbergh M, Gunkel DJ (2023) ChatGPT: deconstructing the debate and moving it forward. AI & Soc. https://doi.org/10.1007/s00146-023-01710-4

DARPA (2016) Broad agency announcement: explainable artificial intelligence (XAI). DARPA-BAA-16-53, Defense Advanced Research Projects Agency

Dennett DC (1978) AI as philosophy and as psychology. In: Ringle M (ed) Philosophical perspectives on artificial intelligence. Humanities Press, New York, pp 57–80

Dreyfus HL (1965) Alchemy and artificial intelligence. Rand Corporation Paper, Santa Monica

Floridi L, Cowls J (2019) A unified framework of five principles for AI in society. Harvard Data Sci Rev 1(1):1–14. https://doi.org/10.1162/99608f92.8cd550d1

Fodor JA (1975) The language of thought. Harvard University Press, Cambridge

Fodor JA (1983) The modularity of mind: an essay on faculty psychology. MIT Press, Cambridge

Fodor JA (1987) Psychosemantics. MIT Press, Cambridge

Friston KJ (2012) The history of the future of the Bayesian brain. Neuroimage 62(2):1230–1233. https://doi.org/10.1016/j.neuroimage.2011.10.004

Gunning D, Aha D (2019) DARPA's explainable artificial intelligence (XAI) program. AI Mag 40(2):44–58. https://doi.org/10.1609/aimag.v40i2.2850

Haugland J (1980) Programs, causal powers, and intentionality. Behav Brain Sci 3(3):432–433

Haugland J (1993) Artificial intelligence. The very idea. MIT Press, Cambridge

Hetherington SC (2011) How to know: a practicalist conception of knowledge. Wiley-Blackwell, Malden

Hofstadter DR (1995) The ineradicable ELIZA effect and its dangers. In: Fluid concepts & creative analogies: computer models of the fundamental mechanisms of thought. Basic Books, New York, pp 155–168

Weizenbaum J (1976) Computer power and human reason: from judgment to calculation. W. H. Freeman and Company, New York

LeCun Y, Bengio Y, Hinton G (2015) Deep learning. Nature 521(7553):436–444

Legg S, Hutter M (2007) Universal intelligence: a definition of machine intelligence. Mind Mach 17:391–444. https://doi.org/10.1007/s11023-007-9079-x

Lieto A (2021) Cognitive design for artificial mind. Taylor & Francis, Milton Park

McCarthy J, Minsky M, Rochester N, Shannon C (2006) A proposal for the Dartmouth summer research project on artificial intelligence, August 31, 1955. AI Mag 27(4):12–14. https://doi.org/10.1609/aimag.v27i4.1904
McCulloch WS, Pitts W (1943) A logical calculus of the ideas immanent in nervous activity. Bull Math Biophys 5(4):115–133
McGeer V (2018) XV—intelligent capacities. Proc Aristot Soc 118(3):347–376. https://doi.org/10.1093/arisoc/aoy017
Nielsen K (2002) The concept of tacit knowledge—a critique. Outlines Crit Pract Stud 4(2):3–17
Nilsson NJ (2009) The quest for artificial intelligence: a history of ideas and achievements. Cambridge University Press, Cambridge
Open Ended Learning Team, Stooke A, Mahajan A, Barros C, Deck C, Bauer J, Sygnowski J, Trebacz M, Jaderberg M, Mathieu M, McAleese N, Bradley-Schmieg N, Wong N, Porcel N, Raileanu R, Hughes-Fitt S, Dalibard V, Czarnecki WM (2021) Open-ended learning leads to generally capable agents. ArXiv:2107.12808. https://doi.org/10.48550/arXiv.2107.12808
Pennachin C, Goertzel B (2007) Contemporary approaches to artificial general intelligence. In: Artificial general intelligence. Springer, Berlin
Polanyi M, Sen A (2009) The tacit dimension. University of Chicago Press, Chicago
Putnam H (1967) The nature of mental states. Originally published as "Psychological Predicates". In: Captain WH, Merrill DD (eds) Art, mind and religion. University of Pittsburgh Press, Pittsburgh, pp 37–48. Reprinted in Mind, language and reality, philosophical papers 2. 429–440. Cambridge: Cambridge University Press, 1975
Putnam H (1988) Representation and reality. MIT Press, Cambridge
Rabern B, Ball D (2018) The science of meaning: essays on the metatheory of natural language semantics. Oxford University Press, Oxford
Rosenblatt F (1958) The perceptron: a probabilistic model for information storage and organization in the brain. Psychol Rev 65(6):386–408. https://doi.org/10.1037/h0042519
Rudi Studer V, Benjamins R, Fensel D (1998) Knowledge engineering: principles and methods. Data Knowl Eng 25(1–2):161–197. https://doi.org/10.1016/S0169-023X(97)00056-6
Rumelhart DE, McClelland JL (1986) On learning the past tenses of English verbs. In: Parallel distributed processing: explorations in the microstructure of cognition, Psychological and psychological and biological models, vol 2. MIT Press, Cambridge, pp 216–271
Russell SJ, Norvig P (2016) Artificial intelligence: a modern approach. Pearson Education Limited, Harlow
Ryle G (2009) The concept of mind, 60th Anniversary ed. 1949. Abingdon/New York: Routledge/Oxon
Schumann N, Yaoli D (2021) Grenzgänge: Von Menschen zu smarten Maschinen – und zurück? In: Strasser A, Sohst W, Stapelfeldt R, Stepec K (eds) Künstliche Intelligenz—Die große Verheißung. Xenomoi Verlag, Berlin, pp 37–62
Schwitzgebel E (2022) The nature of belief from a philosophical perspective, with theoretical and methodological implications for psychology and cognitive science. Front Psychol 13:947664. https://doi.org/10.3389/fpsyg.2022.947664
Searle JR (1980) Minds, brains, and programs. Behavi Brain Sci 3(3):417–457
Searle JR (1984) Intentionality and its place in nature. Dialectica 38(2/3):87–99
Searle JR (2002) Why I am not a property dualist. J Conscious Stud 9(12):57–64
Searle JR (2007) Biological naturalism. In: Velmans M, Schneider S (eds) The Blackwell companion to consciousness. Blackwell, Oxford, pp 327–336
Sellars W (1956) Empiricism and the philosophy of mind. Minn Stud Philos Sci 1:253–329
Shannon C (1948) The mathematical theory of communication. Bell Syst Tech J 27:379–423, 623–656
Silver D, Hubert T, Schrittwieser J, Antonoglou I, Lai M, Guez A, Lanctot M, Sifre L, Kumaran D, Graepel T, Lillicrap TP, Simonyan K, Hassabis D (2018) A general reinforcement learning algorithm that masters chess, shogi, and go through self-play. Science 362:1140–1144

Stanley J, Williamson T (2001) Knowing how. J Philos 98(8):411–444. https://doi.org/10.2307/2678403

Thagard P (2005) Mind: introduction to cognitive science. MIT Press, Cambridge

Von Foerster H (ed) (1951) Cybernetics: circular causal and feedback mechanisms in biological and social systems. Transactions of the seventh conference, Josiah Macy, Jr. Foundation, New York

Wiener N (1985) Cybernetics: or control and communication in the animal and the machine. MIT Press, Cambridge

Chapter 3
A Novel Space for Reframing Mind-Machine Relations

Digitalization and AI technologies are reshaping the world and the way we live in it. This process, described by Floridi as the fourth revolution in human self-understanding, signifies that we are increasingly viewing ourselves as "information organisms" that are "mutually connected and embedded in an informational environment (the infosphere), which we share with other informational agents, both natural and artificial, that also process information logically and autonomously".[1] Developments in computer science and ICTs construct infospheres of information processing and transfer, which are shared by humans and other information artifacts. But how can interconnected information artifacts take part in intersubjective practices? To develop a discussion of this question, this chapter focuses on how digitization processes and semantic technologies offer available resources and environments for both humans and machines. After clarifying why data models and information structures can be used to represent knowledge, it will elaborate on how the internet and semantic technologies have provided a shared knowledge base across domains for networked applications and users. The practice of semantic technologies enhances the interaction and integration of cyberspace and actual space, providing new possibilities for the action and interaction of the information processing agents in it, whether natural or artificial.

[1] Floridi, *The 4th Revolution: How the infosphere is Reshaping Human Reality*, 94. The first three revolutions in human self-understanding are the Copernican revolution, which reveals that humans are not at the center of the universe, the Darwinian revolution, which states that humans are not at the center of the biological kingdom, and the Freudian or neuroscientific revolution, which demonstrates that consciousness is not transparent to oneself. Cf. 87–90.

Y. Du, *Machines in Minds*, Studies in Applied Philosophy, Epistemology and Rational Ethics 78, https://doi.org/10.1007/978-3-032-15842-0_3

3.1 Data, Information, Knowledge

Acquiring knowledge is a characteristic of human intelligence, while information processing agents have the function of processing digital information. What is the relationship between knowledge, information and data? A common model used to represent their relation is the Data-Information-Knowledge (DIK) model.[2] These three terms are distinguished from a functional perspective: Data, as basic facts, are generally considered to be context-free and meaningless; information is structured data that can be used in specific contexts and has corresponding meaning; knowledge, at a more abstract level, organizes and judges information.[3] DIK model can be used to illustrate both the functional hierarchy of information processing in a cognitive subject and the hierarchy of knowledge management in the universal domain. There are different extended versions of the DIK model, in some of which the top layer is action, in others it is shared understanding, and in the more general version of DIKW, wisdom is added to the top layer.[4] Zeleny relates each layer in the DIKW model to a level of knowing, where data is interpreted as knowing nothing, information as knowing what, knowledge as knowing how, and wisdom as knowing why.[5] I'll next explain how this model provides a relational inter-definition of these three concepts and show how this approach is rooted in logical atomism that presupposes basic descriptive facts, the problem with it, and the improvement of it.

3.1.1 *Data and Facts*

We live in a digital society that is full of and constantly generating data, which record, quantify and influence human activities and experiences. The term data was used first in mathematics to express "a fact given as the basis for calculation in mathematical problems" and was introduced into computer science in 1946 to present "transmittable and storable information by which computer operations are

---

[2] Chaim Zins, "Conceptual Approaches for Defining Data, Information, and Knowledge," *Journal of the American Society for Information Science and Technology* 58, no. 4 (2007): 480.

[3] Rowley briefly summarizes data, information, and knowledge in the DIK model: "*Data* can be characterized as being discrete, objective facts or observations, which are unorganized and unprocessed and therefore have no meaning or value because of the lack of context and interpretation. *Information* is described as organized or structured data, which has been processed in such a way that the information now has relevance for a specific purpose or context, and is therefore meaningful, valuable, useful and relevant. *Knowledge* is generally agreed to be an elusive concept which is difficult to define. Knowledge is typically defined with reference to information." Jennifer Rowley and Richard Hartley, *Organizing Knowledge: An Introduction to Managing Access to Information* (London: Routledge, 2016), 6.

[4] David Williams, "Models, Metaphors and Symbols for Information and Knowledge Systems," *Journal of Entrepreneurship, Management and Innovation* 10 (2014): 98.

[5] Jennifer Rowley, "The Wisdom Hierarchy: Representations of the DIKW Hierarchy," *Journal of Information Science* 33, no. 2 (2007): 167.

performed".[6] On the one hand, data are the basic units or symbols of structured information. On the other hand, data refer to the actual electronic transmissions and storage in physical systems.[7]

Both as units and as facts, the concept of data makes a commitment to the facts in the tradition of logical atomism, i.e. atomic facts directly obtained by the senses.[8] Such facts are commonly considered to be immediately perceived without reasoning. This understanding is etymologically in line with the meaning of the word data in Latin, which means something that is "given".[9] The epistemological presupposition here is that sensory data can be directly acquired as the most basic factual unit, but they are passively perceived and presented without further manipulation. Data can therefore be understood as evidence of physical reality, while processing data is presenting physical reality in a neutral mechanism.

In data modeling-based cognitive science, such data processing mechanism can be performed by both humans and computers. It is worth noting that according to the construction of the universal Turing machine, a program is also a kind of data for the machine that reads and executes it. It is a special kind of data that is encoded in such a way that it can manipulate other data. In this case, the process of organizing data is the process of organizing operational machines. However, problems arise when data processing mechanisms and data are treated as pre-given. As we have seen in the representational computational theory of mind, the syntax-driven computing machine, pre-given processing mechanisms cannot account for how the mind understands its relationship to the world. Machine learning and data-driven models shed light on how mechanisms for processing data may lead to technological bias and are not neutral.

In data science, "raw data" often need to be "cleaned" before it can be used for further analysis, which is the process of correcting inaccurate, incomplete, redundant, corrupt, or other data records. Mere data processing lacks an informative context and needs to be interpreted in a further explanatory framework. This indicates

---

[6] Harper Douglas, "Etymology of data," *Online Etymology Dictionary*, accessed July 14, 2022, https://www.etymonline.com/word/data.

[7] In the later case, data can be understood as "one or more kinds of energy waves or particles (light, heat, sound, force, electromagnetic) selected by a conscious organism or intelligent agent on the basis of a preexisting frame or inferential mechanism in the organism or agent". Zins, "Conceptual Approaches for Defining Data, Information, and Knowledge," 481ff.

[8] In the logical atomism proposed by Russell, knowledge about reality is composed of sentences, which in turn consist of proper names and predicates referring to logical atoms. Logical atoms are elementary certainties that cannot be decomposed any further. They are regarded as basic data acquired not through reasoning, but through the senses, or common sense. As Russell writes, "our investigation starts from what may be called 'data', by which I mean matters of common knowledge, vague, complex, inexact, as common knowledge always is, but yet somehow commanding our assent as on the whole and in some interpretation pretty certainly true". Bertrand Russell, *Our Knowledge of the external World as a field of scientific method in Philosophy* (Chicago: Open Court Publishing Co., 1914), 66.

[9] The word data comes from the plural of the Latin word "datum", the past participle of dare, meaning "thing given" or "a fact given or granted". Douglas, "Etymology of data."

that directly collected data need to be fixed in accordance with its purpose to become usable and interpretable information. In this sense, data can also be considered as raw material for information, while information is interpretable data.

3.1.2 *Information and Reality*

In comparison to the concept of data, which contains a commitment to the logical facts of physical reality, the concept of information as an old term reflects on the relationship between logical forms and reality at an ontological level. Capurro provides an etymological examination of the concept of information, beginning with ancient Greece, noting that the term information originated from the exploration of the relationship between form and matter.[10] He argues that both form and matter are categories of thought rather than ontological reality, and that information as a process of "in-formation" is therefore a logical abstraction of form.[11] Therefore, the concept of information can neither be unilaterally attributed to the category of matter nor to the category of form, but it should be considered as a logical category. In other words, the term information contains a logical abstraction of the process through which matter is shaped into form. Yet this understanding of the relationship between information, matter and form has changed in the modern era. In its everyday use, the meaning of information has changed from "giving a (substantial) form to matter" to the modern understanding of conveying some kind of message to someone.[12] This shift is described briefly by Capurro and Hjørland as "from matter to mind".[13] It was not until the rise of computer technology and information theory in the twentieth century that the term information was transformed from an everyday word to a term used in theory.[14]

As mentioned in Sect. 2.2.1, in Shannon's information theory, information is defined by mathematical methods that allow syntactic information to be effectively separated from its semantic content. However, the scope of legitimate applications of purely syntactically defined information theory is unclear, as different research areas study different objects of reality. The definition of information involves the commitments of a particular discipline to its research objects. For example, in

[10] Rafael Capurro, *Information. Ein Beitrag zur etymologischen und ideengeschichtlichen Begründung des Informationsbegriffs* (München: Saur Verlag, 1978), 17ff.

[11] Ibid., 18, 282.

[12] Rafeal Capurro and Birger Hjørland, "The Concept of Information," *Annual Review of Information science and Technology* 37 (2003): 354.

[13] Ibid. Also see John D. Peters, "Information: Notes toward a critical history," *Journal of Communication Inquiry* 12 (1988): 13. Peters notes that Descartes saw the role of "ideas" as the "forms of thought", conveying information to the brain that could be used as raw material for knowledge. That is, information, including the way in which material forms take shape, is no longer considered to be directly recognizable, but mediated through ideas.

[14] Capurro and Hjørland, "The Concept of Information," 354.

biology, the term information is used to refer to the relationship between two entities, such as the causal relationship between genetic and developmental processes, or the relationship between an organism and its environment.[15] In this case, the term information is a generalization of specific relations of materials. Whether syntactic information can consistently describe reality is questioned by the so-called Capurro trilemma.[16] Briefly, if information deals with the same references, then the objects in different domains lose their qualitative differences. If the information refers to something similar or analogous, then the original "prototype" of the analogy needs to be accounted for. And if the information is about different references, then the definition of information is ambiguous. This trilemma indicates the difficulties of a unified theory of information. Some argue that information can be reduced to uniform concepts or to models, while others argue that the concept of information is probably "a network of logically interdependent, but mutually irreducible, concepts".[17] To achieve a standardized and workable definition, information cannot to be defined syntactically only, but its epistemic and normative implications need to be emphasized as well.

Floridi provides a definition of semantic information as syntactically well-formed, meaningful, and truthful data.[18] In his approach, data are understood as relational entities, while information represents the structural features and semantic dimensions of data clusters. Data, either as given facts or forms of given facts, can be used for structuring. Thus, data structures can be treated here as "constraining affordances" that provide structural space for semantic information.[19] Semantic information can then be described as a "*data + queries*" model that works as data structure at the dataset level and has semantics when used by information processing agents.[20] Since semantics deals with meaning and truth, Floridi adds truthfulness to a general definition of semantic information to ensure structured data refers to facts that can be semantically determined to be true.[21] This definition of semantic information concerns the relationship between information and reality. Floridi outlines three dimensions of how information relates to reality:

> Information can be viewed from three perspectives: information *as* reality (e.g. as patterns of physical signals, which are neither true nor false), also known as *environmental*

---

[15] Georg Toepfer, *Historisches Wörterbuch der Biologie. Geschichte und Theorie der biologischen Grundbegriffe*. Band 2. (Berlin: Springer, 2011), 181f.

[16] Capurro and Hjørland, "The Concept of Information", 365.

[17] Luciano Floridi, *The Philosophy of Information* (Oxford: Oxford University Press, 2011), 33.

[18] The investigation of semantic information can be traced back at least to the semantic information theory proposed by Carnap and Bar-Hillel for computing the amount of semantic information in sentences, which can be viewed as an extension of Shannon's theory of information. Floridi notes that semantic information based on this theory can only be defined as well-formed meaningful data, lacking a characterization of its factuality. Ibid., 109ff.

[19] Ibid., 87.

[20] Ibid., 183ff.

[21] Ibid., 86. Truthfulness qualifies semantic information whose truth value can be determined, while false semantic information is classified as pseudo-information.

information; information *about* reality (semantic information, alethically qualifiable); and information *for* reality (instructions, like genetic information, algorithms, orders, or recipes).[22]

First, the notion of information *as* reality is comparable to data as the given form of physical reality. Such information is a direct record or evidence of physical reality and undertakes a commitment to physical reality. But it does not yet have semantics. Second, the notion of information *about* reality is information with semantic relations. In this case information is the description or evaluation of a given fact that can be taken as a proposition. Third, the notion of information *for* reality is a theorization of reality, such as steps or algorithms for doing something. This kind of information can usually be recorded in a complex way. In short, information *as* reality is essentially a part of reality, information *about* reality is a logical expression that can describe reality, and information *for* reality is a theorization that allows the construction of reality in logical space.

Along this line, Floridi proposes an "informational structural realism" that acknowledges "the existence of a mind-independent reality addressed by, and constraining knowledge".[23] This reality is composed of "structured objects that are neither substantial nor material [...] but cohering clusters of data, not in the alphanumeric sense of the word, but in an equally common sense of differences *de re*, i.e. mind-independent, concrete, relational points of lack of uniformity".[24] Reality is understood here as a structured relationship presented through information.

Informational structural realism assumes two minimal ontological commitments, one to the structural properties of reality and the other to structural objects.[25] Floridi argues that the first commitment replies to the question "what can we know?", while the second replies to the question "what can we justifiably assume to be in the external world".[26] The first commitment acknowledges that reality has structural properties. Based on this most basic commitment, reality presented through the structure of information can be recognized. In other words, it assumes that "the ultimate nature of reality is informational".[27] The second commitment allows for the structuring of the objects presented by information so that the first commitment can be justified in epistemic terms.[28] By assuming these two commitments, informational structural realism differs from digital ontology. Digital ontology holds that the ultimate nature of reality is numerical or that the universe is a computational system,

---

[22] Ibid., 31.

[23] Ibid., 340.

[24] Ibid.

[25] Ibid. The ontological commitment can be understood as a commitment to the logical structure of reality, and I will return to this point and explicate its role for the linguistic capacities of rational agents in Sect. 4.2.

[26] Ibid.

[27] Ibid., 341.

[28] Ibid. He writes: "The second commitment, in favor of structural relata, is justified by epistemic reasons. We are allowed to commit ourselves ontologically to whatever minimal conception of objects is useful to make sense of our first commitment in favor of structures." (340)

whereas the ontological commitment of informational structural realism holds only that reality can be presented structurally by information.[29] That is, informational structural realism explores defensible knowledge about reality from an epistemological standpoint. With numerical features, information as a structuring tool provides information processing agents with knowledge and theory that can be tried and tested. He writes:

> Structural objects work epistemologically like constraining affordances: they allow or invite certain epistemic constructs (they are affordances for inforgs like us, who elaborate them) and resist or impede some others (they are constraints for the inforgs), depending on the interaction with, and the nature of, the informational organisms that process them.[30]

Structural objects can be used to construct theories that can be tested and to acquire knowledge of reality. Knowledge and theories obtained from structural objects are information *for* reality, which mediates and regulates the relationship between information processing agents and reality. Such agents are not merely receivers or senders of information, but "embodied and embedded, creative agents, who interact with reality, shape and build it, Plato's artisans, writers not just readers, players not audience, in short *systems designers*".[31] This philosophy of information offers the possibility of a situated cognitive modelling of information agents and environments, rather than pre-given information processing mechanisms and given data in reality.

3.1.3 Knowledge and Authorities

In epistemological discussions about what knowledge is, a classic but arguable view is that knowledge is a justified true belief, whereby a person knows that *p*, meaning that *p* is true, and that person believes that *p* is true and can justify it.[32] This account emphasizes the direct relationship between knowledge and the knowers' particular cognitive states, i.e., beliefs. In this tradition, according to Neuser's analysis of the conception of knowledge in different epochs of Western culture, the inseparable relationship between knowledge and the knowing subject is the dominant doctrine in modern philosophy, where the knower is the authority for the source and

[29] Floridi notes that digital ontology, also called digital metaphysics, can be traced back to Konrad Zuse. He sums up four specific theses of digital ontology: "(1) The nature of the physical universe (time, space and every entity and process in space-time) is ultimately discrete [...] (2) The physical universe can be adequately modelled by discrete values like the integers; (3) The evolution (state transitions) of the physical universe is computable as the output of a (presumably short) algorithm; and (4) The laws governing the physical universe are entirely deterministic." Luciano Floridi, "Against Digital Ontology," *Synthese* 168 (2009): 152f.

[30] Luciano Floridi, *The Philosophy of Information*, 340.

[31] Ibid., 208.

[32] Cf. Edmund L. Gettier, "Is Justified True Belief Knowledge?" *Analysis* 23, no. 6 (1963): 121ff.

justification of knowledge.[33] Yet, with the spread of information technologies in the twentieth century, the notion of knowledge has changed structurally with the transformation of social systems into knowledge society, in which knowledge has become a social resource that can be managed, organized, and transferred. He argues that the current knowledge society can be understood as a post-traditional era of the modern tradition.[34] In this era transition, the authority of knowledge shifts from the subject of theory to the procedure of organizing knowledge with the participation of social practitioners. The process of knowledge constitution extends from subjective thinking to the organization of knowledge in social practices in which individual actions are engaged. Therefore, information in the media and the internet, by virtue of being available as resources to support practice, can also be understood as knowledge.[35]

Against this background, we can better see why models such as DIK have emerged in knowledge management, treating knowledge as an abstract level of data and information, while the latter are the carriers and mediums of knowledge. In the DIK model, knowledge as an abstraction level for information processing also has regard to "knowing how". Rowley explains knowing how in general as tacit knowledge, mainly embodied in empirical judgments that incorporate personal beliefs, perspectives, and values.[36] In contrast, explicit knowledge can be recorded and communicated publicly. This distinction between knowing how and knowing that emphasizes the difference between personally experienced knowledge and publicly communicated knowledge. To clarify the relationship between knowledge involving individual experience and information processing in general form, Zins distinguishes between subjective knowledge and universal knowledge.[37] The former is represented by the individual cognitive agent's state, while the latter can be recorded and represented in the universal domain. Accordingly, he proposes different applications of the DIK model to these two domains:

---

[33] He suggests that Western culture can be divided into different epochs according to the development of conceptual systems in response to the questions of what knowledge is and how it is defended. Ancient, medieval, and modern eras are the traditional phases of three epochs. Correspondingly, the transitional periods of Late Antiquity, the Renaissance, and the contemporary era can be called post-traditional phases. Wolfgang Neuser, *Wissen begreifen. Zur Selbstorganisation von Erfahrung, Handlung und Begriff* (Wiesbaden: Springer VS, 2013), 11.

[34] Significantly, he considers that the modern tradition of subject-based concepts of knowledge ends with the German classical conceptualism. The period from the nineteenth century to the present is its post-traditional phase. Ibid., 43ff.

[35] Neuser argues that information on the Web can be understood as knowledge in two ways. One is from a kind of realist viewpoint, where information can be assumed to be knowledge of objective events. The other, which he promotes, understands the foundations of knowledge as the interrelationship between concepts, experiences, and actions. According to the latter, the criterion of one's knowing is to be able to act on a plan based on knowledge. Knowledge in this sense is a kind of self-organized information resource. Ibid., 48.

[36] Rowley and Hartley, *Organizing Knowledge: An Introduction to Managing Access to Information,* 7.

[37] Zins, "Conceptual Approaches for Defining Data, Information, and Knowledge," 486.

> D-I-K in the subjective domain. In the subjective domain, data are sensory stimuli, which we perceive through our sense. Information is the meaning of these sensory stimuli [...] Knowledge is thought in the individual's mind, which is characterized by the individual's justifiable belief that it is true.
>
> D-I-K in the universal domain. In the universal domain, data, information and knowledge are human artifacts. They are represented by empirical signs [...] Meaning, in the objective domain data are sets of signs that represent empirical stimuli or perceptions, information is a set of signs, which represent empirical knowledge, and knowledge is a set of signs that represent the meaning (or the content) of thoughts that the individual justifiably believes that they are true.[38]

This distinction represents two different ways of understanding data, information, and knowledge. In the subjective domain, data, information, and knowledge correspond to subjective sensory stimuli, experiential perceptions, and justified true beliefs, respectively. Individual experience is understood as being constituted from these three layers. By contrast, in the universal domain, data, information, and knowledge correspond to sets of symbols that record and store experience, such as books, reports, and logs. In this case, data, information, and knowledge as symbol sets are artifacts that are recorded and transmitted. Both for the subjective and universal domains, the DIK model is a functional hierarchy of data, information, and knowledge. Whereas it characterizes the functional hierarchy of cognitive subjects in the subjective domain, in the universal domain it is the functional hierarchy of symbolic systems that can be shared and interpreted. Neuser distinguishes similarly between individual and general knowledge systems in the information technologies-supported knowledge society.[39] Knowledge in both systems involves concepts, experiences, and actions, but the individual knowledge system is self-organizing, while the general knowledge system is self-stabilizing.[40] In the interplay between the two systems, the constitution of knowledge is shifted between the authoritative subject organizing knowledge and the individual user participating in the management and stabilization of knowledge. Here, knowledge sharing is not reducible to information communication in interconnected general knowledge systems, as it involves procedures of formulating protocols and norms.

Since the birth of artificial intelligence research, knowledge engineering has been one of the main approaches to simulate intelligent performance such as storing knowledge, representing knowledge, and reasoning about knowledge.[41] Knowledge engineering focuses on architecting an intelligent system capable of reasoning and making decisions that mimic the way humans acquire and use knowledge. Semantic networks are commonly used to represent propositional knowledge in early research, where semantics is understood as the study of "the relations between signs or

[38] Ibid., 487.

[39] He argues that general knowledge can be stored not only in information media but also managed in information processing-based settings, such as social networks or robotic systems, while the latter is knowledge based on experiences of subjects. Neuser, *Wissen begreifen*, 253.

[40] Ibid., 263ff.

[41] Russell and Norvig, *Artificial Intelligence: A Modern Approach*, 307f.

symbols and what they denote or mean".[42] Semantic networks are motivated by 1960s theories about human memory, which assume that associative memory is network-based for extracting information.[43] Since semantic relations are used to model associative memory, a semantic network combines both knowledge storage and knowledge inference functions in a single mechanism.[44] This simple and efficient inference mechanisms are the main advantage of the semantic networks. The most common relation is of the form "IS-A", while inheritance is a common inference rule that allows a subclass to inherit the properties of the superclass. However, there are many difficulties encountered in semantic networks design. First, it is difficult for semantic networks to represent complex sentences, such as those containing relative clauses.[45] This problem is based on the limitations of the first-order predicate logic used by simply semantic nets. Second, it is a challenge to connect more semantic networks together, because the labels of nodes or links in different simple semantic nets may be distinct. The lack of standardization limits the ability to reason about contextual transformations.

A more critical problem lies in the challenge of how knowledge-based intelligent systems can respond to an ever-changing world. The original frame problem was formulated by McCarthy and Hayes, which is essentially a question of how an agent keeps track of persisting facts when reasoning about change.[46] From a cognitive point of view, the frame theory of data structures in intelligent systems can be compared with schema or schemata studies in cognitive psychology.[47] These studies assume that the human information processing system uses frame-like structures, meaning that knowledge is not acquired and updated simply by the constant accumulation of believed propositions, but is integrated and adapted continuously by an intrinsically internal structure.[48] Haugland acutely indicates that in questioning how a representational schema matches a new situation, one already presupposes that the structure of the representational schema is internally fixed, and so it becomes problematic how the added information affects the existing structure.[49] This presupposition distinguishes knowledge acquisition from representation, assuming that

---

[42] William A. Woods, *What's in a Link: Foundations for Semantic Networks* (Cambridge, MA: Bolt, Beranek and Newman, 1975), 9.

[43] Nilsson, *The Quest for Artificial Intelligence: A History of Ideas and Achievements*, 205.

[44] Woods, *What's in a Link: Foundations for Semantic Networks*, 14.

[45] Ibid., 37–66.

[46] John McCarthy & Patrick J. Hayes, "Some philosophical problems from the standpoint of artificial intelligence," in *Machine intelligence,* vol. 4, eds. Bernard Meltzer and Donald Michie (Edinburgh: Edinburgh University Press, 1969), 487.

[47] Paul R. Thagard, "Frames, knowledge, and inference," *Synthese* 61, no.2 (1984): 245.

[48] Fodor formulizes the frame problem as: "How, then, does the machine's program determine which beliefs the robot ought to re-evaluate given that it has embarked upon some or other course of action?" Jerry A. Fodor, *Modularity of Mind* (Cambridge, MA: MIT Press, 1983), 114.

[49] John Haugeland, "An overview of the frame problem," in *The robot's dilemma: the frame problem in artificial intelligence* ed. Zenon. W. Pylyshyn (Norwood: Ablex Publishing Corporation, 1987), 77–93. He indicates that "the challenge is not how to decide for each fact whether it matters, but rather how to avoid that decision for almost every bit of knowledge".

information is represented in fixed schemas, while the acquired sense data changes with the environment. The frame problem arises from the tension between the inherent syntax of representational schemas and the changing semantics of sensory data.

From a technical point of view, the frame problem deals with the problem of how to construct a formal system that deals with complex and changing conditions. It is thus not directly about representational contents, but about modeling and updating the relationships of the contents.[50] In terms of effectively designing data structures for knowledge representation, Minsky proposes using frames to represent knowledge. He conceives of frames as a kind of data structure "representing a stereotyped situation", analogous to the way people situationally recall information.[51] In contrast to simple semantic networks, frames are object-oriented data structures that allow a frame to be described by different information and to establish multiple relations with other frames. Object-oriented conceptualization allows data structures to represent exceptions and differences in the same concept in a flexible manner.[52] The frame data structure can be written in frame language, which can classify and order unstructured data according to object-oriented conceptualization by using classifiers, a kind of inference engine that was later adopted by ontology languages.

The frame problem reveals the difficulties of computational representational theory. Frame-based data structures provide a procedural approach to structuring data and their relationships. This avoids switching between different data models and instead places various semantic relations in a consistent conceptual framework. However, object-oriented conceptualisation involves a deeper question of whether there are and how to define conceptually fundamental objects.[53] How and who can ensure that data models are both object-oriented and conceptually consistent? This brings into play the specification of conceptualization methods.

The explicit specification of conceptualization can be made by information ontologies.[54] Drawing on the philosophical term ontology for the study of existence, information ontologies are introduced to formally define basic entities and their

---

[50] Cf. Lars-Erik, Janlert, "Modeling change: the frame problem," in *The robot's dilemma: the frame problem in artificial intelligence*, ed. Zenon. W. Pylyshyn (Norwood: Ablex Publishing Corporation, 1987), 65–76.

[51] He stresses that each frame has "several kinds of information. Some of this information is about how to use the frame. Some is about what one can expect to happen next. Some is about to do if these expectations are not confirmed." Marvin Minsky, "A Framework for Representing Knowledge," in *The Psychology of Computer Vision*, ed. Patrick Winston (New York: McGraw-Hill, 1975), 212.

[52] Cf. Ben Coppin, *Artificial Intelligence Illuminated* (London: Jones and Bartlett, 2004), 35.

[53] Hayes notes that frame theory involves a deeper ontological problem because "to use frames is to make a certain kind of assumption about what entities shall be assumed to exist in the world." Patrick Hayes, "The Logic of Frames," in *Frame Conceptions and Text Understanding* (Berlin: De Gruyter, 1980), 4.

[54] Thomas R. Gruber, "A Translation Approach to Portable Ontology Specifications," *Knowledge Acquisition* 5, no. 2 (1993): 199.

classes, properties, etc. in a domain.[55] Taxonomy is a typical classification method used to stratify the basic items of a domain.[56] But fixed classifications usually limit the scope of uses of the described relations and lack the flexibility to translate knowledge across domains. To organize and represent knowledge from different domains, a general-purpose ontology is needed to ensure consistency when concepts from different domains are translated to each other in a shared universe of discourse.[57] Constructing information ontologies can be seen as an engineering application of the philosophical question of "what is there?".[58] Information ontologies can thus be understood as the abstraction of data models, ensuring logical consistency between concepts on the one hand, and committing to the validity of their referents on the other. Gruber suggests that in constructing an information ontology, an informational realist stance can be taken, i.e., that "what 'exists' is precisely that which can be represented".[59] This is the basic commitment underlying information ontologies to the relationship between data models and reality. Data models mediated by an information ontologie, thus, do not relate to reality in a directly referential way, but rather conceptualize committed realities in terms of agreements or protocols in accordance with basic logical relations provided by ontologies.

Information ontologies are usually specified by ontology languages with classifiers for defining and categorising concepts and thereby constructing and organising knowledge. A current widely used ontology language is the Web Ontology Language, which will be detailed in the next subchapter in terms of its role in Semantic Web technologies. We will see how the standardization of conceptualization with the help of ontology languages and corresponding data models facilitates the evolution of the internet. With the support of Semantic Web technologies, what mediates data and reality is no longer extra information ontologies, but rather interlinked data and metadata themselves, whose conceptualization is specified by the practices of Internet users. I will argue that cyberspace, supported by semantic technologies, allows information and reality to shape each other into shared web resources and environments available to information processing agents.

---

[55] Thomas R. Grube, "Ontology," in Encyclopedia of Database Systems eds. Ling Liu and M. Tamer Özsu (Boston: Springer, 2009), 1963.

[56] In addition to taxonomies, topologies are also basic classification methods. They differ in that typology conceptually classifies representative and ideal items, while taxonomy classifies observed items based on experience. Accordingly, typology is typically applied to the social sciences, while taxonomy is used in disciplines such as biology and in cluster analysis. Cf. Kenneth D. Bailey, *Typologies and Taxonomies. An Introduction to Classification Techniques* (Thousand Oaks, CA: Sage, 1994), 4–10.

[57] Gruber, "A Translation Approach to Portable Ontology Specifications," 203.

[58] Quine considers ontology to be the study of what there is. I will unfold his account of ontological issues and ontological commitments from the logical and linguistic philosophical viewpoints in Sect. 4.2.

[59] Gruber, "A Translation Approach to Portable Ontology Specifications," 200.

3.2 Semantic (Web) Technologies in Digital Social Practices

The increasing digitization of everyday practices offers new possibilities for knowledge-based, data-driven AI. Digitization is the process of converting information recording the world we live in into digital format for computer processing.[60] Compared to continuous analog raw materials, data in discrete digital formats can be more easily stored, disseminated, and accessed. This is because digital formats are used to simulate different materials in analog form, such as text, images, sound, etc. Today the process of digitization extends from specific areas of expertise to various aspects of everyday life, including but not limited to the development of the internet, the Internet of Things, and the networking of people as users with each other.[61] The information environment created by today's digital ICTs is described by Floridi as the infosphere:

> *Minimally*, infosphere denotes the whole informational environment constituted by all informational entities, their properties, interactions, processes, and mutual relations. [...] *Maximally*, infosphere is a concept that can also be used as synonymous with reality, once we interpret the latter informationally. In this case, the suggestion is that what is real is informational and what is informational is real.[62]

According to this informational structural realistic view, the objects in the ICTs environment, and their properties, relations are defined through ICTs. An important subregion of the infosphere constituted by ICTs is cyberspace for ordinary users, which is commonly used to refer to the internet.[63] The popularity of the internet has enabled the creation of a wider range of knowledge engineering, triggering an information explosion of everyday data. Roughly speaking, the internet is a global network that interconnects regional networks. The idea of connecting different computers and regional networks began in the 1960s with the network packet switching method.[64] Packet switching is a communications technology that "divides the input flow of information into small segments, or packets, of data which move through the network in a manner similar to the handling of mail but at immensely higher speeds."[65] A packet is the basic unit of data transmission, and it allows data to be transferred between regional networks using addresses like postal packages. The transmission of packets requires standardization of how data are grouped,

---

[60] Ronald Deckert, *Digitalisierung und Industrie 4.0. Technologischer Wandel und individuelle Weiterentwicklung* (Springer, Wiesbaden 2019), 2.

[61] Ibid., 9.

[62] Floridi, *The 4th Revolution: How the inforsphere is Reshaping Human Reality*, 41.

[63] Floridi emphasizes, "[Infosphere] is an environment comparable to, but different from, cyberspace, which is only one of its sub-regions, as it were, since the infosphere also includes offline and analogue spaces of information." Ibid.

[64] Cf. Lawrence G. Roberts, "The Evolution of Packet Switching," *Proceeding of the IEEE* 66, no. 11 (1978): 1307ff.

[65] Ibid., 1307.

addressed, and received. Standardization is an important aspect of ICTs and the internet.

The standardization of information communication involves a series of communication protocols. Protocols are components of the internet infrastructure in addition to hardware such as servers and devices. Current data communications on the internet can be described by the Open Systems Interconnection (OSI) model developed by the International Organization for Standardization (ISO). This conceptual model details the protocols and functions of data communication at different logical layers:

As presented in Fig. 3.1, the OSI model clearly describes the communication protocols involved in the process of data transmission over the internet in seven logical layers. The first layer is the physical hardware processing layer, which transforms raw data and physical signals. The second layer is the data link layer, which provides data transmission between nodes. The third layer is the network layer that links the different networks, serves as a medium to connect the different sub-networks, and controls the operation of the sub-networks. The fourth layer is the transport layer, which ensures that the information is delivered in sequence and with integrity. The fifth is the session layer, which establishes information exchange between resources and nodes. The sixth layer is the presentation layer, which converts the data format into a readable format at the application level. The seventh layer is the application layer for end user access to information. In addition to these seven layers, in some versions there is an additional eighth layer that represents a

	Layer	Protocol Data Unit	Function	Examples
Host layers	**7 Application**	Data	High-level application programming interfaces, including resource sharing, remote file access	HTTP, FTP
	6 Presentation		Translation of data between a networking service and an application; including character encoding, data compression and encryption/decryption	XDR, TLS, SSL
	5 Session		Managing communication sessions, i.e., continuous exchange of information in the form of multiple back-and forth transmission between two nodes	PPTP, SAP, L2TP
	4 Transport	Segment, Datagram	Reliable transmission of data segments between points on a network, including segmentation, acknowledgement and multiplexing	TCP, UDP, SPX
Media layers	**3 Network**	Packet	Structuring and managing a multi-node network, including addressing, routing and traffic control	IPv4, IPv6, IPX, ICMP
	2 Data link	Frame	Reliable transmission of data frames between two nodes connected by a physical layer	IEEE.802.3, IEEE. 802.11
	1 Physical	Bit, Symbol	Transmission and reception of raw bit streams over a physical medium	DSL, Bluetooth

Fig. 3.1 The Open Systems Interconnection model (Adapted from https://en.wikipedia.org/wiki/OSI_model)

"user" in a personal sense, a ninth layer that represents an organization, and a tenth layer to represent government or legal regulations. Each layer in the OSI model is a logical abstraction of the function of protocols within it, and each layer serves the layer above it. This hierarchy makes it possible to track how data are received or sent at each logical layer. This model shows that data transmission over the internet is always protocol dependent. Data transfer on the internet is not an end-to-end network but involves the entire architecture of information communication.

3.2.1 Semantic Web Architecture and Tools

The World Wide Web (WWW), or simply the web, is the most widely used system for accessing the internet today. The concept of the web was invented by Bernes-Lee in 1990 to create a common information space where people could share information with each other and work together.[66] For this purpose, the web provides a specific way to organize and share information on the internet. The core technologies of the web include the Hypertext Transfer Protocol (HTTP) the protocol for users to access web resources, the Hypertext Markup Language (HTML) to describe web resources published on websites, and Uniform Resource Locators (URLs) to identify web resources.[67] Hypertexts are texts or documents that are interconnected by hyperlinks, while hyperlinks are links used to access or refer to web resources. These core technologies enable website owners to publish their information on their websites and allow other users to access that information.

In the early stage of the Web or the so-called Web 1.0, the transfer of information between users was one-way. Ordinary users could only read information from a website, but could not contribute to or modify it.[68] To have a more flexible way of sharing web resources, Bernes-Lee envisions a "Semantic Web" that is not only read-only but read-write, and more importantly, a web that not only connects people but also enables computers to communicate and share information. He writes:

> I have a dream for the Web [...] and it has two parts. In the first part, the Web becomes a much more powerful means for collaboration between people. [...] In the second part of the

---

[66] Tim Berners-Lee, "The World Wide Web: A very short personal history," May 7, 1998, http://www.w3.org/People/Berners-Lee/ShortHistory.html.

[67] Sareh Aghaei, Mohammad Ali Nematbakhsh and Hadi Khosravi Farsan, "Evolution of the World Wide Web: From Web 1.0 to Web 4.0," *International Journal of Web & Semantic Technology* 3, no. 1 (2012): 3.

[68] In contrast to Web 1.0, the term Web 2.0 is used to describe a human-centered Web in which there are more participants and interactions. The main technologies of Web 2.0 include blogs, Really Simple Syndication (RSS), tagging, etc. These technologies allow users to contribute content online in a flexible way of using information. This also falls under Semantic Web development, as I will detail below. But the Semantic Web is sometimes also taken as a synonym for the term Web 3.0, which is used to describe the web that focuses on the relationships between data and is not limited to the relationships between users. Ibid., 5–7. Also see, Luciano Floridi, "Web 2.0 vs. the Semantic Web: A philosophical Assessment," *Episteme* 6, no. 1(2009): 25–37.

> dream, collaborations extend to computers. Machine become capable of analyzing all the data on the Web – the content, links, and transactions between people and computers. A "Semantic Web", which should make this possible, has yet to emerge, but when it does, the day-to-day mechanisms of trade, bureaucracy, and our daily lives will be handled by machines talking to machines, leaving humans to provide the inspiration and intuition. The "intelligent agents" people have touted for ages will finally materialize.[69]

The Semantic Web is proposed to enhance the availability of web resources for machines. The traditional web can be seen as a huge repository of interconnected Hypertexts, in which documents can be transmitted and presented by machines, but they can only be read and interpreted by humans. By contrast, the Semantic Web is designed to make web resources machine-readable, allowing computers to access, read, and process resources on the web just like human users. Thus, unlike the traditional Web, consisting of files that are readable by human users, the Semantic Web is a Web of data that is readable and processable by computers. One of the purposes of Semantic Web development is to prepare available resources and environments for machine agents that process web resources like humans. The development of the Semantic Web project is led by the World Wide Web Consortium (W3C), which provides a detailed description of the Semantic Web:

> The Semantic Web provides a common framework that allows data to be shared and reused across application, enterprise, and community boundaries. [...]
>
> The Semantic Web is about two things. It is about common formats for integration and combination of data drawn from diverse sources, where on the original Web mainly concentrated on the interchange of documents. It is also about language for recording how the data relates to real world objects. That allows a person, or a machine, to start off in one database, and then move through an unending set of databases which are connected not by wires but by being about the same thing.[70]

By this definition, the Semantic Web provides a common framework for sharing web resources. This framework gives a common format and syntax to data from different sources so that the data can be processed by machines, and ensures the consistency of the referential relationships between the data and related objects. In short, the Semantic Web can be understood as a set of technologies and standards for structuring and integrating data from different databases according to the way people describe objects in the real world. The core technologies and standards of the Semantic Web can be conceptually represented in a layered structure by the Semantic Web Stack[71]:

Similar to the ISO model, the Semantic Web Stack architecture describes the operations, protocols, languages, and the functions required for different logical

[69] Tim Berners-Lee and Mark Fischetti, *Weaving the Web: The Original Design and Ultimate Destiny of the World Wide Web by Its Inventor* (New York: Harper, 2000), 157f.

[70] "W3C Semantic Web Activity," W3C, last modified December 11, 2013, https://www.w3.org/2001/sw/.

[71] Aurona Gerber, Alta van der Merwe, and Andries Barnard, "A Functional Semantic Web Architecture," in *The Semantic Web: Research and Applications*, eds. Sean Bechhofer et al. (Berlin, Heidelberg: Springer, 2008), 273.

levels of data from the physical level to the application level.[72] As shown in Fig. 3.2, the architecture of the Semantic Web can be divided into four groups, the base layer in red, the syntactic layer in orange, the semantic or framework layer in gray, and the application layer in green. A detailed explanation is provided below.

The most fundamental base layer has two main components. One component is Unicode, the international standard used to encode character sets that allow various writing systems to be encoded into information that can be processed. The other is the Uniform Resource Identifier (URI), which is a standardized unique sequence of characters to identify various resources on the internet. That is, identifiers are used to access or refer to web resources, and URIs refer to web resources like names in natural language.[73] Notably, to avoid ambiguity between resources with the same name, each individual name specified by a URI has a namespace (NS) to ensure that all identifiers are unique. This identification method ensures that web resources can have a unique reference.

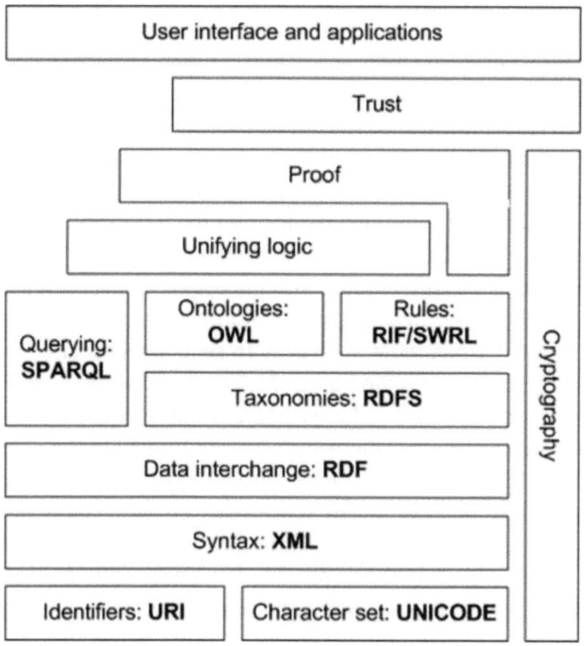

Fig. 3.2 Semantic Web Stack (Adapted from https://en.wikipedia.org/wiki/File:Semantic-web-stack.png?uselang=en#filelinks)

[72] Ibid., 276.

[73] Harry Halpin, *Social Semantics. The Search for Meaning on the Web* (New York: Springer, 2013), 43. Both URIs and URLs were first used to identify hypertext and documents. A URL is a special kind of URI, an identifier that can indicate how to access a web resource.

Next, the identified resources acquire syntax through the Extended Markup Language (XML). XML provides a syntactic form to mark up and distinguish web resources, so that the resources can be encoded in a format that is readable not only by humans but also by machines. As a markup language, XML is designed to carry and integrate data, not just display it. In addition, XML provides only syntax yet has no established labels. This brings more flexibility and compatibility for adding and extending data. Since the role of syntax is to form structural relationships between symbols, XML involves only structuring data with syntax, or marking up the document structure with an XML schema language. It does not represent or process any semantics.

Semantic relationships of web resources are described by the Resource Description Framework (RDF), which shares a unified data model with XML. RDF describes the semantic relationships of Web resources through the interpretative model between metadata and data. Metadata are literally defined as data about data. For example, if the content of a document is data, its author, date, format, and related tags and labels are metadata. Metadata can thus be explicated as "indications about the nature of some other (usually primary) data. They describe properties such as location, format, updating, availability, usage restrictions, and so forth."[74] The interpretative relationship between metadata and data is relative, and metadata are also data. Metadata can also be structured and constitute referential relationships between data. In fact, RDF was originally designed as a data model for metadata to provide related information for web resources.[75] Web resources can be stored and retrieved in a relational form through the form of the RDF triple. In past knowledge bases, since the conceptualization of object-oriented data models is relatively stable while data about instances is dynamic, the conceptual models and instances of data were suggested to be stored separately.[76] In contrast, by structuring data and metadata relations in RDF triple, there is no need to separate data and their models. The task of RDF is therefore to provide a mechanism for describing web resources by means of interpretive relationships between metadata and data. It provides a general formal semantic description mechanism that is not limited to a specific domain.

Taken together, URIs, XML and RDF provide a set of semi-structures for data integration and interchange. With this semi-structure, data can be uniquely identified on the one hand and interrelated with other web resources on the other. The RDF schema (RDFS) is designed to provide the basic reasoning for the resources described by RDF, pre-attributing their types and properties.[77] RDFS and RDF form

[74] Floridi, *The Philosophy of Information*, 88.

[75] "Resource Description Framework (RDF): Concept and ABSTRACT Syntax," W3C Recommendation, last modified February 10, 2004, https://www.w3.org/TR/rdf-concepts/.

[76] According to Gruber, the ontology of conceptualization provisions about data models can be placed in a T-Box, a set of terminological components. The counterpart is the A-Box, a set of assertion components about instances. Gruber, "A Translation Approach to Portable Ontology Specifications," 206.

[77] Aghaei, Nematbakhsh, and Farsani, "Evolution of the World Wide Web: From Web 1.0 to Web 4.0," 7.

the RDF family, providing formal semantic descriptions of data resources. Formal semantics is concerned with how symbolic structures convey meaning, as I will elaborate in depth in Sect. 4.1. Although symbolic structures do not contain semantics per se, they provide mutually interpretable relationships that can be read and processed by machines. The symbolic structures available to interpret and be interpreted provide the agent with the basic patterns of understanding. I will discuss in Sect. 4.2 how such basic patterns of interpretation afford humans the possibility to interpret whether a nonhuman agent is thinking.

The Ontology Web Language (OWL) is an ontology language for the Semantic Web. OWL is "designed to represent rich and complex knowledge about thing, group of things, and relations between things".[78] As mentioned in the previous section, ontology languages provide common vocabularies for describing data from different domains, enabling data to be exchanged and shared in a cross-domain universe of discourse. OWL embeds the semi-structure of XML/RDF in its syntax to allow richer integration and interoperability of data resources between different domains. For reasoning rules that cannot be presented in OWL directly, other languages need to be developed to code the syntax. The Rule Interchange Format (RIF) is composed of these languages. In contrast to RDFS, OWL, RIF is developed to cover rule-based logical reasoning.

SPARQL is a standard RDF query language, consisting of the SPARQL Protocol and the RDF Query Language. SPARQL allows accessing and retrieving triple patterns of RDF data schemas. Since both RDFS and OWL are syntactically based on the RDF model, SPARQL can also be used directly for querying constructed data taxonomies and ontologies.

The top layers of the Semantic Web stack architecture, the green part of Fig. 5, are expected technologies that have not yet been implemented. Since the basic idea of the Semantic Web architecture is to structure and merge heterogeneous data sources for the user interface and applications, it is necessary to ensure that data sources are accurate and trustworthy. The logic and proof layers, and the trust layer are designed to provide applications with structured data that is quality assured and can be used for reasoning. Eventually, trustworthy, accurate and structured data sources are available to the user interface and applications.

Above is a short introduction to the Semantic Web architecture and related technologies. The original design of the Semantic Web architecture and its actual development did not exactly match, especially after December 2013, when the W3C subsumed Semantic Web activity under linked data activity.[79] The purpose of the linked data activity, like the Semantic Web, is to build the web of machine-processable data, not just the web of documents. This is the focus of the next section. Although the project agenda changed, the core technologies of the Semantic

[78] "Web Ontology Language," W3C, last modified December 11, 2012, https://www.w3.org/2001/sw/wiki/OWL.

[79] "W3C Data Activity Building the Web of Data," W3C, accessed June 07, 2021, https://www.w3.org/2013/data/.

Web remain the same. Before moving on to the next section, there are a few more points worth stressing about the conception of the Semantic Web and its architecture.

The first is the relationship between the Semantic Web and AI development. As already indicated, the purpose of the Semantic Web is to provide machine-readable web resources. Machine readability is not about processing information according to established laws. XML and the RDF family constitute the core Semantic Web technologies that allow semi-structured data to be processed flexibly by machines. The Semantic Web is in this sense "a descendant of semantic networks from classical artificial intelligence, where nodes are resources and arcs are links."[80] In other words, the Semantic Web can be regarded as a general-purpose database that provides readable and reasonable web resources for knowledge-based AI.[81] The development of the Semantic Web has turned knowledge representation into a global knowledge engineering project.

Second, the semi-structured syntax provided by XML and RDF allows the Semantic Web to organize data in a way that is not only object-oriented, but also relational. Berners-Lee points out that the RDF model is more general than the usual entity-relationship model because the "entities" in the RDF data model are "nodes" whose semantics are not given but are determined by their relationships with other nodes.[82] That is, the relationships between data are the first class of objects in the RDF data model, which can be created by anyone and identified by URIs. The advantage of this is that technically anyone can add information about the data on the web, constructing a more general and open model of entities and their relationships. Information transfer on the web is no longer one-way from providers to end users, but it is disseminated and modified by numerous users.

Third, it is important to note that the Semantic Web does not provide a consistent reasoning system. Berners-Lee suggests that to connect data from different domains in an efficient way, the idea of global consistency must first be abandoned.[83] This is because the goal of the Semantic Web is to present various aspects of real life not to theorize the lifeworld into logical reasoning. Semantic Web technologies thus identify and structure web resources without fixing their definitions, allowing data to describe each other, and keeping data open for operation.

To put it briefly, Semantic Web technologies provide a common relational data structure that allows data from different data models to be interlinked, web resources and their relationships to be manipulated flexibly. This means that web resources can be structured by semantic technologies into syntactic forms that are interrelated,

---

[80] Halpin, *Social Semantics. The Search for Meaning on the Web*, 56.

[81] Tim Berners-Lee, "What the Semantic Web can represent," Design Issues, last modified September 17, 1998. https://www.w3.org/DesignIssues/RDFnot.html.

[82] Ibid. Berners-Lee notes that the Semantic Web resembles the relational database model. Relational databases refer to databases founded on a relational model of data, initially organizing and adding data in rows and columns in the form of table cells. Since the Semantic Web is designed to link data from different data models, it provides a common data structure in an open form like a relational database and allows the data to be accessible by a structured query language.

[83] Ibid.

and their semantics are flexible. With core semantic technologies such as RDF, web resources and their relations can be put into special syntactic structures with formal semantics. By adding descriptions about data such as labels for web content, web users can generate metadata that adds relationships and context between web resources. In this way, the Semantic Web as a networked database can provide machine information processing agents with readable and inferable resources.

3.2.2 Knowledge Engineering for Structuring Data and Metadata

In 2013, the W3C attributed the activity principle of the Semantic Web to Linked Data, a core concept that had been introduced by Berners-Lee has introduced in the design issues about the Semantic Web. The Semantic Web is the web of data, not the web of documents. And data are not isolated. They are interlinked and have traceable relationships. As he puts it:

> The Semantic Web isn't just about putting data on the web. It is about making links, so that a person or machine can explore the web of data. With linked data when you have some of it, you can find other, related, data.[84]

It is the connections between data that give them semantic relationships that refer to each other rather than in isolation. As described in the previous section, semantic technologies make web resources and their relationships machine-readable and processable. Relationships between web resources are established through links, which serve as connections between resources, allowing the starting resource to refer to the ending resource. Valid links in the Semantic Web should be identified by URIs. The construction of the Semantic Web follows the "principle of linking", i.e., "any resource can be linked to another resource identified by a URI. No resource is an island, and the relationships between resources are captured by the linking, transforming lone resources into a Web".[85] For this purpose, Bernes-Lee specifies the following rules for making the web grow:

1. Use URIs as names for things
2. Use HTTP URIs so that people can look up those names
3. When someone looks up a URI, provide useful information, using the standards (RDF*, SPARQL)
4. Include links to other URIs. so that they can discover more things.[86]

These rules guide users to link data using standardized technologies so that web resources are not stored in isolation but are linked in the form of common data

[84] Tim Berners-Lee, "Linked Data," *Design Issues*, last modified June 18, 2009. https://www.w3.org/DesignIssues/LinkedData.html.

[85] Halpin, *Social Semantics. The Search for Meaning on the Web*, 45.

[86] Berners-Lee, "Linked Data."

model that can be retrieved across the web. Meanwhile, Bernes-Lee proposes the concept of Linked Open Data (LOD) to encourage linked data published through open licenses, allowing conceptual systems from different domains to be standardized on a large scale.[87] The LOD project has been led by the W3C since 2007. It is defined as follows:

> Linked Open Data defines a vision of globally accessible and linked data on the internet based on the RDF standards of the semantic web. LOD is often thought of as a virtual data cloud where anyone can access any data they are authorized to see and may also add to any data without disturbing the original data source. This provides an open environment where data can be created, connected and consumed on internet scale.[88]

LOD aims to provide an open and globally accessible cyberspace where any authorized data and the relationships between them can be published and added to. The basic idea behind creating LOD is that "data has more value if it can be connected to other data", and data in this context means "any structured web-based information".[89] This means that the data in LOD are information on the web structured with semantic technologies such as URI, XML/RDF data model. In this open web space, the value of data lies in its connection to other data.

The amount of linked data has increased rapidly in the dozen years. The development and current landscape of LOD can be found at LOD-Cloud.[90] A short overview of the development of LOD, in 2007 there were only 12 datasets in LOD, while by 2021 there are already 1301 datasets containing at least 16,283 links between datasets.[91] A dataset is a structured collection of data. As the largest dataset in LOD, the DBpedia Community Project extracts data from information created by various Wikimedia projects. Among them are over 400 million facts describing 3.7 million things extracted from the English version of Wikipedia alone.[92] These data are structured and integrated into a large-scale and multilingual knowledge base to support other applications. The DBpedia Community Project maintains the real-time knowledge base, keeping it up to date with changes to the Wikipedia page.[93] In other words, DBpedia provides a real-time knowledge base covering a wide range of topics for numerous applications, providing efficient interconnected hub for

---

[87] Ibid.

[88] "Linked Open Data," W3C, last modified June 7, 2010. https://www.w3.org/egov/wiki/Linked_Open_Data.

[89] Ibid.

[90] The Linked Open Data Cloud, accessed January 15, 2022. https://lod-cloud.net/.

[91] Ibid.

[92] "The DBpedia knowledge bases that are extracted from the other 110 Wikipedia editions together consist of 1.46 billion facts and describe 10 million additional things. […] The project publishes regular releases of all DBpedia knowledge bases for download and provides SPARQL query access to 14 out of the 111 language editions via a global network of local DBpedia chapters. […] DBpedia sets 27 million RDF links pointing into over 30 external data sources and thus enables data from these sources to be used together with DBpedia data." Jens Lehmann et al., "DBpedia—A Large-Scale, Multilingual Knowledge Base Extracted from Wikipedia," *Semantic Web* (2015): 167.

[93] Ibid.

organizing, sharing, and searching information. In this way, Wikipedia is not only an encyclopedia for human users, but also an encyclopedia or knowledge base for networked artificial intelligent systems and agents. This is also one of the major purposes of LOD.

Linked Data shares roughly the same core technologies as the Semantic Web, using semantic technologies such as URI, XML, RDF, OWL. As mentioned in the previous section, RDF serves as a data model for storing and interchanging data in the form of RDF triples. An RDF triple can express a statement consisting of a Subject-Predicate-Object structure. According to W3C:

An ***RDF triple*** contains three components:

- the ***subject***, which is an *RDF URI reference* or a *blank node*
- the ***predicate***, which is an *RDF URI reference*
- the ***object***, which is an *RDF URI reference*, a *literal* or a *blank node*

An RDF triple is conventionally written in the order subject, predicate, object.

The predicate is also known as the ***property*** of the triple.[94]As can be seen above, each part of the triple can be individually identified and processed by an RDF URI reference, which is a Unicode sequence that produces a valid URI sequence. The predicate or property in the triple must be a URI, because it represents a link between two resources. And if the resources of the subject or object in the triple is not already identified by a URI, a blank node can be used instead. Besides, values such as strings, numbers, and dates, which are generally used as objects describing resources in assertions, are represented as literal.[95] While any statement expressed in RDF can be represented as triple, a set of RDF triples is called an RDF graph. In an RDF graph, the subject is the starting resource, the predicate is the link, and the object is the ending resource. An RDF graph can be visualized "as a node and directed-arc diagram, in which each triple is represented as a node-arc-node link".[96] Starting from the basic triple structure, more complex triple structures can be built and extended by RDF. Because a triple structure can itself be assigned a URI, it can then be linked to other URIs as part of a new triple. This resembles the process of forming sentences from clauses in natural language. In this way, the RDF model allows web resources to form complex and large interconnected graphs. The RDF data model is thus also known as graph-based data model.

In the form of RDF triples, data from different sources can be integrated into statements that can be expressed in subject-predicate object form. And the flexible way of linking RDF graph structures allows the relationships between data not to be

[94] "Resource Description Framework (RDF): Concepts and Abstract Syntax, W3C Recommendation (2004)," W3C, February 10, 2004, https://www.w3.org/TR/rdf-concepts/. In version 2014, URIs are replaced by IRIs (Internationalized Resource Identifiers). "URIs and IRIs: IRIs are a generalization of URIs [RFC3986] that permits a wider range of Unicode characters. Every absolute URI and URL is an IRI, but not every IRI is an URI." "RDF 1.1 Concepts and Abstract Syntax," W3C, February 25, 2014. https://www.w3.org/TR/rdf11-concepts/.

[95] "RDF 1.1 Concepts and Abstract Syntax". https://www.w3.org/TR/rdf11-concepts/.

[96] Ibid.

limited to given descriptions, but rather depends on how the statements represented by the RDF graph are used, in relation to other statements. RDF thus provides semantics for semi-structured data in a formal and pragmatic sense:

> An overarching goal in the use of RDF is to be able to automatically merge useful information from multiple sources to form a larger collection that is still coherent and useful. As a starting point for this merging, all the information is conveyed in the same simple style, subject predicate-object triples, as described above. To keep the information coherent, however, we need more than just a standard syntax; we also need agreement about the semantics of these triples. [...]
>
> An RDF graph is "true" exactly when all the triples in it are "true". These notions, and others, are specified with mathematical precision in the RDF Semantics document.[97]

RDF graphs are constructed in a compositional way based on triples. When all triple components are taken to be true, that RDF graph is also considered true. This holistic semantic feature results from the fact that the RDF data model treats relations as its primary class. This relation-oriented data structure is more flexible than a merely object-oriented data model in terms of cross-domain interoperability.[98] In a merely object-oriented model, objects are treated as primary classes, and information about objects needs to be described at the meta-level, e.g., through taxonomic ontologies to classify them. In this case, metadata and data are in different class levels and are separated. On the contrary, relations are the primary class in an RDF graph, linking different objects. Once the links are formed, a triple structure is constructed that allows the data to describe each other. This means that a set of data can be described by other data as well as used to describe other data. In this case, metadata themselves are data.

Furthermore, RDF allows for complex descriptions, because the basic triple structure can in turn be used as structural components to form more complex statements. In this way, from the basic triples it is possible to assemble large RDF graphs in which the data are structured as a combination of mutual descriptions. This is the key to linked data based on Semantic Web technologies. The data model provided by semantic technologies are summarized by Heath and Bizer as follows:

- **A unifying data model**. Linked Data relies on RDF as a single, unifying data model. By providing for the globally unique identification of entities and by allowing different schemata to be used in parallel to represent data, the RDF data model has been especially designed for the use case of global data sharing. In contrast, the other methods for publishing data on the Web rely on a wide variety of different data models, and the resulting heterogeneity needs to be bridged in the integration process.
- **A standardized data access mechanism.** Linked Data commits itself to a specific pattern of using the HTTP protocol. This agreement allows data sources to

97 "RDF 1.1 Prime," W3C Working Group Note, June 24, 2014. https://www.w3.org/TR/rdf11-primer/.

98 Tim, Berners-Lee, "Relational Databases on the Semantic Web", Design Issues, September 1998, last modified August 27, 2009. https://www.w3.org/DesignIssues/RDB-RDF.html.

be accessed using generic data browsers and enables the complete data space to be crawled by search engines. In contrast, Web APIs are accessed using different proprietary interfaces.

- **Hyperlink-based data discovery.** By using URIs as global identifiers for entities, Linked Data allows hyperlinks to be set between entities in different data sources. These data links connect all Linked Data into a single global data space and enable Linked Data applications to discover new data sources at run-time. In contrast, Web APIs as well as data dumps in proprietary formats remain isolated data islands.
- **Self-descriptive data.** Linked Data eases the integration of data from different sources by relying on shared vocabularies, making the definitions of these vocabularies retrievable, and by allowing terms from different vocabularies to be connected to each other by vocabulary links.[99]

This overview highlights linked data as a process of standardizing data models across the global Web, with the aim of making data from different sources more operational. The process of standardization is the process of establishing and unifying internet protocols. Linked data are gradually referred to each other in the standardization and become metadata for each other. Accordingly, the shared vocabularies complete their self-referencing and self-description. The achievement of linked data needs to motivate more users to participate, because mutual depictions of data are based on the metadata generated by the large amount of data.

In the process of digitalization, the term "Big Data" describes the vast amount of data generated as the lifeworld is constantly recorded in data formats, big in terms of volume, variety, velocity, etc.[100] These data include unstructured "raw data" that normally needs to be "cleaned" before it can be used for further analysis, but it is worth noting that the generation and aggregation of these data are already embedded in socio-digital practices.[101] That is, the data generated by digital social practices are not neutral, but can be considered as a kind of socio-technical device for further interpretations or narratives. The implied norms and value preferences of these data can be unfolded by semantic technologies to classify and merge the data by adding explicit semantic relations:

> Mixing data from different sources, and doing so at scale, can be hard. It requires a layer within the architecture that adds meaning and interoperability to the data; in other words, it needs a semantic layer. The use of URI references as identifiers and RDF-encoded

[99] Tom Heath and Christian Bizer, *Linked Data: Evolving the Web into a Global Data Space* (San Rafael: Morgan & Claypool, 2011). HTML Version. http://linkeddatabook.com/editions/1.0/.

[100] In addition to three V's, Kitchen adds exhaustiveness, high resolution, relationality and flexibility to characterise big data. Rob Kitchin, *The Data Revolution: Big Data, Open Data, Data Infrastructures and their Consequences* (London: Sage, 2014), 99ff.

[101] Kitchen argues that "no data are pre-analytic, or objective and independent elements" and uses the term "data assemblages" to characterize data as complex socio-technical apparatuses. Ibid., 50ff.

> ontologies to create a semantification application is a key component of the Big Data Europe Platform […][102]

By identifying with URIs, and semi-structuring data with RDF models, data can form new relationships that are created in their use. Halpin argues that the Semantic Web is thus endowed with "social semantics" in terms of the public and social aspects of language.[103] The use of language is not private, but a shared practice involving multiple interlocutors. This is because linguistic activity as the establishment and practice of using referential relations, is a social activity. The process of naming and constructing relationships between web resources resembles the actual process of constructing semantic relationships. Halpin writes:

> The *social semantics position* states that since the Web, including the Semantic Web, is a form of language, as a language exists as a mechanism for co-ordination among multiple agents, then the meaning of a URI is the use of the URI by a community of agents, a use that can be *traced* by the computational sense of the URI ranging across tags, queries, and whatever representations are hosted at the URI.[104]

By identifying web resources with URIs and assigning namespaces, resources acquire semi-structured syntax that construct relationships as atomic facts. This process also makes the web a logical space for structured information. Moreover, from a social semantics viewpoint, representation is no longer a mapping of established reality within a system. Knowledge representation is the internalization of socially engineered knowledge in individuals. Constructing semantic relations is not about mapping the set of relations that already exist in the reality, but about constructing and structuring reality with information. That is, knowledge representations of fixed relations are inadequate to portray semantic relations, and such representations are based on the ability to construct semantic relations. Constructing semi-structured semantic relations can be seen as an ability to make formal judgements that are provisional semi-categorical rather than categorical. The practice of this ability is in accordance with the activity of users acquiring and generating information while using the web. This is because the logical constructs of the web are influenced by the language use and behavior of a wide variety of users, such as using search engines, adding, or modifying tags, etc.

The process of modifying data and their states in the web by surfing the internet is compared by Halpin to a "language game", a term coined by Wittgenstein to illustrate that the relationship between language and reality is not simply a representational one through naming, but is linked in different contexts through language use and action.[105] Language games are varied, ranging from pointing to something and describing it, to giving commands through words. The meaning of a linguistic expression then depends on the context in which it is used. The rules of language

[102] "W3C and BIG DATA," W3C, October 19, 2016, https://www.w3.org/blog/2016/10/w3c-and-big-data/.

[103] Halpin, *Social Semantics. The Search for Meaning on the Web*, 187.

[104] Ibid.

[105] Wittgenstein, *Philosophical Investigation*, 5. §7.

use are neither fixed nor arbitrary, but are enacted in the interplay between language users and the context. In short, language games are social practices involving language. The process of identifying a resource by keywords is also a kind of language game, just like finding a resource by keywords in a search engine. This act is performed by numerous users. Halpin writes:

> [T]yping in terms into a search engine can be thought of as what Wittgenstein would term a 'language-game,' with success in this game being judged in terms of whether or not a given user can discover a relevant URI via the right keywords. The terms themselves may be ambiguous, but it does not matter if a relevant URI is discovered. In this way, the objective sense of a URI can be considered the search terms that can be used by multiple users to find a particular URI. What a URI means is precisely the set of search terms that leads multiple users to discover the URI in the context of satisfying a particular information need.[106]

The descriptions from different users during the search process determine the meaning of a URI. Search engines can track which URIs are clicked by users and store the data in query logs, which can then be used to improve the ranking of the corresponding search results for the keyword. Users can also make web resources easier to search by adding tags and labels to them. In other words, users are contributing to the structure of the web as they use it, although this engagement in contributing is not recognized by most users. This is because "it *appears* that searching the web using a search engine is a private experience, it is in fact mediated by a vast amount of web pages stored in the search engine's index and the behavior of previous search users".[107] That is to say that a user surfing the web is guided and regulated by the behavior traces of other users. The indirect mutual impact of users in the identification and linking of web resources makes this particular language game an intersubjective process of semantic assignment. This intersubjective language game mirrors the process of semantic assignment and standardization in everyday social practice. I will elaborate in Chap. 5 on how language games in social practices can condition the mutual attribution of mental states by the game participants.

With the development of RDF graph-based Linked Data, search engines have become smarter, such as Google Knowledge Graph project that started in 2012, which aims to enable search engines not only to present search results, but also to explain the meaning of search terms and provide relevant context.[108] In the blog post *Inside the Knowledge Graph: things, not strings*, the main idea is presented as organizing global information and making it universally accessible and useful.[109] On the one hand knowledge graphs undertake conceptualization in a way that is consistent

[106] Halpin, *Social Semantics. The Search for Meaning on the Web*, 191.

[107] Ibid.

[108] Heiko Paulheim, "Knowledge graph refinement: A survey of approaches and evaluation methods," in *Semantic Web* 8 (2017): 490.

[109] The Google Knowledge Graph has accumulated more than 500 billion descriptions about 5 billion things in 2020. Danny Sullivan, "A reintroduction to our Knowledge Graph and knowledge panels," May 20, 2020, https://blog.google/products/search/about-knowledge-graph-and-knowledge-panels/.

with how concepts are used in the real world, and on the other hand, knowledge graphs in their use can provide new relationships to things in the real world. This is because searching for things online and labeling them can affect how they are related and ordered. Knowledge graphs turn the web into a knowledge base for search engines, where relationships between structured data and the world are specified when queried by information processing agents.

As a result, the internet has become a place for generating machine-readable semantic information and for engineering knowledge in the digitization process supported by semantic technologies. What has evolved of the internet is not only the scale, but also the development of exchangeable data models and standards. The structuring and querying of linked data simulate language games in everyday social practices such as naming, labeling, and classifying. Web resources are inter-contextualized and regulated in the online language games of users. Put differently, semantic technologies allow the internet to become an interlinked general knowledge base under construction, which does not follow established rules of reasoning but assigns relationships based on user feedback. The syntactically flexible semi-structure of RDF enables web resources to be endowed with social semantics in their use or, more specifically, as I will explain in Chap. 4, formal semantics based on externalized epistemologies.

Recalling the initial purpose of the Semantic Web to provide machine-readable resources, semantic technologies have greatly contributed to the development of AI applications. This is because Semantic Web technologies provide both human users and artificial agents with readable web resources based on flexible data structures. These resources are accessible as general knowledge by both human and artificial information processing agents. Training on datasets constructed from these resources shared with human users significantly improves the learning outcomes of sub-symbolic AI. Typical examples are large-scale language models that have made significant breakthroughs recently, such as the Generative Pre-trained Transformer (GPT) family developed by OpenAI. Starting with the second generation of GPT, the datasets used for training incorporate corpora that include web resources extracted from Wikipedia, social networks, and other web pages.[110] The datasets used for training contain resources that have been filtered and accessed online by different human users. The third generation of datasets builds on this foundation by expanding the resources crawled from web pages.[111] These datasets are assigned different weights according to their quality, higher weights mean more reliable and more likely to be sampled. I will discuss in detail the training mechanisms of the GPT family in Chap. 6, and the challenges faced by such internet-based AI technologies when engaging in social practices.

---

[110] To better train the language model for unsupervised learning, OpenAI developed a dataset called WebText, which is a dataset of 45 million links crawled from external linked pages that have been liked at least three times on Reddits. Alec Radford, et al. "Language Models are Unsupervised Multitask Learners," *OpenAI blog* 1 (8), 2019, 9.

[111] Tom B. Brown, et al. "Language Models are Few-Shot Learners," *arXiv: 2005.14165v4*. July 22, 2020.

3.3 Linked Cyberspace as a Materialized Logical Space

As semantic technologies enable data to be structured into machine-readable and interoperable resources, information exchange between internet-connected machines is not limited to direct communication between machines, but can be an interplay between them and their shared environment. Cyberspace, as the central domain in the infosphere provided by ICTs, is the shared environment in which the information processing agents in communication are located. The term cyberspace is literally a combination of cybernetics and space. The term has appeared in science fiction and pioneering artworks since the 1960s, and it became widely popular with the rise of the internet in the 1990s.[112] In its original use, the term cyberspace referred specifically to virtual spaces or virtual environments, emphasizing the distinction from everyday reality. The term cyberspace is today used to refer to the environment of computer- and network-mediated experience and communication, as a synonym and metaphor for the internet.[113] The global networking of cyberspace and digital devices lead to the technological vision of ubiquitous computing, the idea of augmenting the present reality with networked information and computing anytime and anywhere.[114] Accordingly, cyberspace is distinct from the offline lifeworld on the one hand, and is on the other hand built on the infrastructure of the real world and conceived as assisting in broadening the experience of the lifeworld. Cyberspace is thus not opposed to reality or an arbitrary extension of it, but rather a digital extension of social practices rooted in normative socio-technical frameworks.

3.3.1 The Order of "Things" in Cyberspace and "Material Intentionality"

Cyberspace includes, in a broad sense, networking infrastructure as well as a variety of connected devices and things that contain embedded sensors and software technologies. These devices and things can be connected to the internet or other networks to form the Internet of Things (IoT), which is a more descriptive term for the technological vision of ubiquitous computing.[115] IoT can be understood as "a concept that encompasses a variety of technologies and research areas that aim to

[112] Cf. Lance Strate, "The varieties of cyberspace: Problems in definition and delimitation," *Western Journal of Communication* 63, no. 3 (1999): 382–83.

[113] Cf. Mark Graham, "Geography/internet: Ethereal alternate dimensions of cyberspace or grounded augmented realities?" *The Geographical Journal* 179, no. 2 (2013): 177–182.

[114] Cf. Klaus Wiegerling, "Ubiquitous Computing und informatische Vernetzung," In *Handbuch Technikphilosophie*, eds. Mathias Gutmann, Klaus Wiegerling and Benjamin Rathgeber (Stuttgart: J.B. Metzler, 2024), 348f.

[115] According to Wiegerling, the difference between the technologies referred to by "ubiquitous computing", "the Internet of Things" and "Ambient Intelligence" is marginal. Ibid., 345.

extend the existing Internet to real-world objects".[116] It attempts to extend the way devices communicate and share information on the internet to real objects, based on networking of addressable interconnected objects with standardized communication protocols.[117] Due to the development of sensor hardware, embedded systems, and other technologies, the IoT can be extended to everyday environments where connected devices are used. Smart Home, Internet of Medical Things, and Industrial IoT are the main application areas of IoT. IoT is an overlap of things in cyberspace and things in real space. On the one hand, IoT turns application scenarios into a hybrid of cyberspace and the real world, providing the basis for information sharing between devices. Things in the IoT can send the data they collected by their sensors to the cloud, the network of centrally processing data, for storage and computation. On the other hand, IoT also enhances the interaction between devices and human users by improving the sharing of information between devices. Interconnected things therefore refer both to objects networked in physical space and to their regulated use in cyberspace. In this sense, it becomes clearer that cyberspace is no longer a virtual space detached from the reality, but an augmentation of it.

The early stages of development of the IoT started with localized environments, dealing only with things in specific domains. The development of the internet and semantic technologies allows the IoT to expand its scope to provide a global, smart environment that connects things from different domains through standardized communication protocols.[118] Regardless of the scale of application, the purpose of the IoT is to connect identifiable things so that things and their states can be represented and regulated uniformly in cyberspace.[119] The network not only regulates the information about things represented in cyberspace, but also their status and role in reality. The things in the IoT are both digital objects in cyberspace and physical objects in actual space. In small-scale scenarios the emphasis is on the information capture capability of the sensors of things or devices, while in large-scale scenarios particular focus is placed on the dependence of the networking on communication protocols. This is to explain IoT in terms of object-centric and internet-centric respectively. The development of IoT is also characterized by the slogan "Internet of Everything", which expects people, processes, data, and things to be linked.[120]

---

[116] Tomas S. Lopez et al., "Adding sense to the Internet of Things An architecture framework for Smart Objective systems," *Pers Ubiquit Comput* 16 (2012): 291.

[117] Ray summarizes several typical definitions of IoT from diverse aspects: (1) IOT is a dynamic global network infrastructure based on interoperable communication protocols and standards; (2) IOT enables things with identities and virtual personalities to operate in smart spaces using intelligent interfaces; (3) IOT is a global infrastructure for the information society implementing advanced services. Cf. Partha P. Ray, "A survey on Internet of Things architectures," *Journal of KING Sud University - Computer and Information Sciences* 30, no. 3 (2018): 293.

[118] Cf. Ioan Szilagyi and Patrice Wira, "Ontologies and Semantic Web for the Internet of Things—a Survey," in *42nd Annual Conference of the IEEE Industrial Electronics Society* (2016): 6949–54.

[119] Depending on its application environment scenarios, IoT can be defined as small-scale and large-scale. Ibid., 6949.

[120] Ronald Deckert, *Digitalisierung und Industrie 4.0. Technologischer Wandel und individuelle Weiterentwicklung,* 15.

Currently, the most important applications of IoT are still in industry. IoT technologies drive the development of the so-called Industry 4.0, which is the fourth industrial revolution after mechanization, electrification and (preliminary) automation.[121] Industry 4.0 stands out for intelligent automation and interconnection. It requires building smart factories based on the industrial IoT as a basis for sharing data on smart things and connecting digitally networked production, producers, and consumers. The main technology for smart factories is cyber-physical systems (CPS), the convergence of computational and physical systems.[122] In this case, cyberspace serves as a local space for integrating computation and physical processes. Both CPS and IoT involve devices embedded in computer systems. While IoT focuses on connecting things through the internet, CPS stresses the integration of physical processes through networks, not necessarily the internet. Early developments in CPS focused on equipping physical objects with identification techniques and sensors, while more recent CPS developments have focused on compatibility evaluation. Thus, the development of CPS is increasingly drawing on IoT to prioritize the connectivity of devices and objects, rather than just developing computing network systems.[123]

The main purpose of interconnecting embedded devices and things is to provide a better user experience. Both the experience of human users with things and the communication between users are enhanced by information communication and semantic technologies. These technological developments have increased the interaction and integration of cyberspace and real space. A new technological environment for user experience has emerged, which is conceptually termed Ambient Intelligence (AmI). Verbeek explains the new technological environment and AmI as follows:

> The miniaturization of electronic equipment and improvements in wireless communication between devices has led to the design of so-called ‘smart environments’. These environments register what happens around them and respond intelligently to this information. The technology at work here is often invisible and is, furthermore, painstakingly coordinated to suit human cognitive processes. That is why it has been named ‘ambient intelligence’. When combined with cognitive and behavioural science, such technologies can also be used to deliberately influence the ideas, intentions, and behaviour of human beings.[124]

Intelligent environments and the devices within them are designed to conform to the behavioral habits and cognitive processes of human users. AmI is intelligent because it provides a technological environment in which both machines and human users can share information, and based on this, allows machines and humans as agents to

---

[121] Ibid., 12.

[122] Baodouin Dafflon, Nejib Moalla, and Yacine Ouzrout, “The challenges, approaches, and used techniques of CPS for manufacturing in Industry 4.0: a literature review,” *The International Journal of Advanced Manufacturing Technology* 113 (2021): 2396.

[123] Jay Lee, Bagheri, Behrad and Kao, Hung-An, “A cyber-physical systems architecture for industry 4.0-based manufacturing systems,” *Manufacturing Letters* 3 (2015): 18.

[124] Peter-Paul Verbeek, “Ambient Intelligence and the Persuasive Technology: The Blurring Boundaries Between Human and Technology,” *Nanoethics* 3 (2009): 231.

regulate and alter their respective states and behavioral patterns. This new technological environment can even influence the ways users interact with each other and affect the patterns of human perception and behavior. In other words, AmI provides intelligent environments and devices that are mutually shaped with human users. The AmI research promotes a human-centered approach, due to its aims to improve the experience of human users in both physical and social level interactions.[125] And one way to improve the user experience is to simplify the complexity of technologies at the application level and blur the boundaries between technology and everyday life. This is a key point of today's smart product design, presenting technological devices as "ready-to-hand" everyday objects.[126]

From hand axes to smartphones, technological development and human habits are mutually constrained and facilitated. Technological environments and social activity scenarios are also merging with each other. Humans modify their own living environment, and the modified technological environment influences the behavioral pattern of humans. The role of the technological environment on human behavior is described by Verbeek as "material intentionality" of technology, which depends on human social intentionality and acts upon it in turn. He writes:

> The active role played by technology does not imply that technologies have intentions like people do, after all they cannot purposefully do anything. Their lack of consciousness does not, however, alter the fact that technologies can have intentions in the original, literal sense of the Latin word *intendere*, which means to 'give direction': technologies give direction to someone's actions or consciousness. [...] Technological mediation can thus be seen as a specific, material form of intentionality. By mediating in the relationship between human and reality, technologies give direction to people's actions and experiences.[127]

Technology has the role of actively directing behavior. Such directing is not attributed to the intentional direction of agents, but rather the material affordance that the environment provides for the possibilities of individual behaviors. The term affordance was initially introduced by Gibson to demonstrate that objects in the environment can provide agents with recognizable possibilities for action.[128] Just as a cave can be perceived for its potential to keep out the rain, or a hand axe can be grasped for its potential to chop, the environment can provide the agent with recognizable clues to possible action, where environmental conditions and the agent's intentional action are intertwined. Intentionality, as meant in the above quote, can then be distributed to the process of interaction between the agent and the environment. In this consideration, the source of intentionality is extended from the intrinsic nature of intelligent agents to the co-constitution of agents and their cognitive ecologies.

---

[125] Ibid., 233.

[126] Ibid., 236.

[127] Ibid., 235.

[128] Gibson introduced the concept of affordance to account for the interface between animals' sensory systems and their environment. Briefly, he argued that the action possibilities offered by objects in the environment can be perceived. Cf. James J. Gibson, *The ecological approach to visual perception* (Boston: Houghton Mifflin, 1979), 127 ff.

Along these lines, networked things and information in smart environments can also afford recognizable possible actions for information processing agents. In particular, the current technological environments formed by ICTs and semantic technologies connect physical objects, integrate cyberspace and actual physical space, expanding the possibilities of user experiences and actions. The technological environments created by today's ICTs and internet technologies is termed by Aydin, González Woge and Verbeek as Active Technological Environments, which not only mediate humans and the world but also "merge with our world, and at the same time have a bidirectional intentional relation with humans."[129] New technological environments are active because they not only afford possibilities for action, but also inform and instruct the ways in which information processing agents act. This bidirectional relationship is not just an interaction between individual agents and things, but involves the co-construction of broader socio-cultural frameworks and technological environments. In the socio-technical systems constituted by socio-cultural frameworks and technological environments, networked things are no longer context-free atomic objects, but interrelated, interoperable affordances. As a result, the new technological environment can provide not only human users but also artificial information processing agents with directions for actions that can be explained by sociocultural frameworks.

3.3.2 A Materialized Logical Space for Information Processing Agents

With the development and increasingly widespread availability of ICTs, the internet and semantic technologies, digitization processes connect and integrate established datasets, the semantic networking being built, and actual physical space into new technological environments. The imitation of the ability to establish semantic relations has shifted from internal knowledge representation to the engineering of a networked environment. Cyberspace supported by semantic technologies provides possibilities for the development of human-centered AI technologies and the reasonable action of technological agents. This can be explained in the following two specific ways.

First, cyberspace is no longer a virtual space in its original sense in opposition to reality, but a means of augmenting reality and extending user experience and action. Smart environments and the things connected therein cover both cyberspace and actual space. Cyberspace emerged in its early days as a virtual space distinct from reality and is now becoming part of everyday experience. The order of things in cyberspace corresponds to the logical space of social practices in the real world. The logical space is neither a subjective psychological space nor a natural physical

[129] Ciano Aydin, Margoth González Woge and Peter-Paul Verbeek, "Technological Environmentality: Conceptualizing Technology as a Mediating Milieu," *Philosophy & Technology* 32 (2019): 336.

space, but an intersubjectively recognized space containing logical objects and normative rules. In the development of ICTs, internet infrastructures and protocols, as well as semantic technologies and standards, data generated in social practices are structured and linked into a materialized logical space. The materialized logical space as cyberspace contains structured data and their links that are committed to objects and their relations in the real-life world. Through the continuous simulation of the real-life world and the modeling of the knowledge relations therein, cyberspace participates in the process of norm-setting in the lifeworld. This means that cyberspace is not merely imitating how social practice occurs, but is the place where social practice occurs. This materialized logical space is not a mapping of semantic relations in the already established lifeworld, but rather serves as a data-driven interface to social practices, structuring the data generated by digital social practices.

Second, the logical space constituted by semantic technologies supported cyberspace provides accessible and available data resources for smart devices and connected things. This offers new possibilities for the development of knowledge-based AI. Unlike classical AI approaches that pursue modeling how knowledge is stored and retrieved internally, the development of the Semantic Web instead begins by modifying the resources and the environment in which intelligent behavior operates. As Berners-Lee emphasizes "the Web was designed as an information space, with the goal that it should be useful not only for human-human communication, but also that machines would be able to participate and help."[130] Human users are embedded in the infosphere as information organisms in the fourth revolution, like organisms living in their biosphere. Interlinked artifacts can also share this resource environment with humans as artificial information processing agents. This means that the starting point of the Semantic Web for intelligence is the interaction of humans and their environment. This interaction results in the adaptation and construction of the environment by human practices, providing the infrastructure for intelligent agents that perform intelligently in this environment. Halpin borrows the concept of "man-machine symbiosis" to indicate that the goal of the Semantic Web is to build an environment in which human and machine agents can cooperate.[131] In an environment supported by semantic technologies, artificial agents do not simulate a presumed intrinsic ability to represent knowledge, but rather they model social knowledge engineering based on the interaction between agents and their shared environment. Both humans and artificial agents share the available resources in the environment and engage in social interactions to enact behavioral patterns and norms. Such knowledge-based AI technologies are thus not human-like, but human social practices-centered.

Together with these two points, cyberspace supported by semantic technologies allows human and artificial agents to share an environment where meaning-making and knowledge engineering are taking place. Increasingly digital social practices generate readable and usable resources for both humans and machines, and the

[130] Tim Berners-Lee, "What the Semantic Web can represent."

[131] Halpin, *Social Semantics. The Search for Meaning on the Web*, 81.

application of AI technologies has expanded from specialized fields to all aspects of daily life. Against this background, the revival of connectionism has contributed machine learning methods based on large artificial neural networks to data science. Combining the respective strengths of symbolic and sub-symbolic AI approaches to develop hybrid AI approaches capable of explainable reasoning and learning has become a growing trend.[132] But this combination is not just a patchwork of existing methods, but rather constructing learning architectures for training on web resources. AI approaches driven by data integrated across the internet supported by semantic technologies do not mimic some intrinsic mechanisms of the human mind anymore, nor do they treat the plasticity of the nervous system as an inherent ability detached from situation. Rather, the modeling of cognitive processes should be extended from within an agent to the process of communication and interaction between the agents and their environment. Intelligence in this understanding would no longer be the ghost in the machine, but rather a specific way of processing information practiced by the agents in their interaction with environments. Thinking by analogy with software, and more specifically mental representations and computational procedures by analogy with data structures and algorithms, should also be extended to the practice of agents accessing data structures as resources in their environment and enacting appropriate information processing procedures.

This extended model is consistent with the view of situated cognition, which emphasizes that knowing and acting cannot be separated during cognitive processes and focuses on the environmental conditions in which cognition occurs, i.e., “the structures of the world and how they constrain and guide human behavior”.[133] The important implication of situated cognition is that the storage and effective retrieval of knowledge depend on how knowledge is accessed in action and perception. This calls for breaking down internal and external spatial metaphors. Actions and their intentions cannot be solely attributed to the internal processes of the actor, but need to be accounted for in terms of the relationship between actors and the situations in which they are embedded. Suchman applies the situated cognitive approach to the study of human-machine interaction, especially to the study of action planning.[134] She argues against the view that task performance is information processing based on a pre-given plan and holds that the plan for task execution is developed when using resources. That is, performing a task can be understood as a non-preplanned and flexible way of acting, where actors need to integrate and use resources to

[132] Marcus suggests “a hybrid, knowledge-driven, reasoning-based approach” to AI development, which should be centered on cognitive models”. Gary Marcus, “The Next Decade in AI: Four Steps Towards Robust Artificial Intelligence,” *arXiv preprint arXiv:2002.06177* (2020), 1.

[133] Donald A. Norman, “Cognition in the Head and in the World: An Introduction to the Special Issue on Situated Action,” *Cognitive Science* 17, no. 1 (1993): 4.

[134] She argues that “actions are always situated in social and material circumstances; the situation is crucial to action’s interpretation”. Lucy A. Suchman, *Human-machine Reconfigurations: Plans and Situated Actions* (Cambridge: Cambridge University Press, 2007), 176.

develop and adjust plans according to their situation.[135] Based on this view, she argues that human-machine interactions resemble human-human interactions, distinguished only by the different resources available to the participants.[136] The resources available to humans are usually richer than in the case of machines. Lacking a shared environment laden with machine-available resources, human-computer interaction is not reciprocal.

Current cyberspace can provide machine-readable resources supported by semantic technologies, and machines that benefit from it can act and interact more appropriately. Machine intelligence based on human social practices is termed by Smart as "Human-Extended Machine Cognition" (HEMC).[137] He suggests that the internet, internet-based multiple artificial intelligent systems, and algorithms can together be treated as the groundwork for extended cognition for both human users and non-human agents.[138] HEMC is based on what Smart proposes as the web-extended mind hypothesis, which states that "the technological and informational elements of the Web can (at least sometimes) serve as part of the mechanistic substrate that realizes human mental states and processes".[139] This hypothesis is an adaptation of the extended mind thesis, which originally holds that resources outside the boundaries of the individual can also partially constitute cognitive processes.[140] Accordingly, the occurrence of cognitive processes and the vehicles of cognitive states cannot be reduced to individuals' brain or body, but rather to a dynamic system consisting of the brain, the body, and the environment in which it is situated. The extended mind thesis is thus considered to be adoptable to explain how the internet as an external technological system extends the human cognitive system.[141] According to the extended mind thesis, a cognitive system can be a coupled system of individual organisms and external resources, so long as the external resources satisfy the criteria of availability, trust, and accessibility.[142] Smart argues that web resources can be evaluated by drawing on these criteria, and once they are

[135] The situated cognitive process is thus a process of resource utilization where action plans should be viewed "as resources for action rather than as controlling structures". Ibid., 183.

[136] For example, "people make use of a rich array of linguistic, nonverbal, and inferential resources in finding the intelligibility of actions and events", whereas machines "rely on a fixed array of sensory inputs, mapped to a predetermined set of internal states and responses". Ibid., 179.

[137] Paul R. Smart, "Extended Cognition and the Internet. A Review of Current Issues of Controversies," *Philosophy and Technology* 30 (2017): 383.

[138] Human agents and artificial agents mutually augment each other in the technological environment, where the information processing of the machine forms part of the cognitive and decision-making system of the human agent, while human agents "form part of the physical machinery for non-biological forms of cognitive processing". Ibid.

[139] Paul R. Smart, "The Web-extended mind," *Metaphilosophy* 43, no. 4 (2012): 447.

[140] Cf. Andy Clark and David Chalmers, "The Extended Mind," *Analysis* 58, no. 1 (1998): 7–19.

[141] Smart, "Extended Cognition and the Internet. A Review of Current Issues of Controversies," 359.

[142] Andy Clark, "Memento's Revenge: The Extended Mind, Extended," in *The Extended Mind*, ed. Richard Menary (Cambridge, MA: MIT Press, 2010), 46.

met, an extended cognitive system containing web resources as constituent material elements can be reached.[143]

To account for the fact that the internet extends the cognitive systems of not only human but also non-human agents, it is important to note the asymmetric cognitive resources of human and nonhuman agents. The original version of the extended mind thesis is concerned with how the mind extends to an organism's surroundings centered on the central nervous system.[144] That is to say, the extended mind thesis presupposes an "un-extended mind" and assumes that the mind can extend outward from the "internal biological realm".[145] Aydin notes that although the extended mind thesis defends externalism by arguing that external matter actively influences internal forms of organization, it is still influenced by the tradition of the primacy of internal space.[146] The internal and external can only be coupled into an extended cognitive system if the utilization of external resources resembles internal cognitive processes. In that case, however, it is difficult for the original extended mind approach to explain how engagement with external matter can become part of cognitive processes that are already self-organized internally. And if it is accepted that cognition is not extended from the inside out, but is mutually shaped by the environment and the information processing agents, then organism-centered cognitive modeling presuppositions are unnecessary. Aydin thereby proposes an artifactual mind thesis, arguing that "our thinking is not pre-given or naturally present in a presumed inside world but that it unfolds itself by virtue of and through objects and artifacts."[147] This thesis can be well compatible with HEMC, since human cognition is here conditioned by objects and especially by the material environment generated by human practices. Smart argues that the environment of human practices expanded by internet technologies not only extends the cognitive systems of human users, but also extends the information processing of computer systems into a networked distributed form.[148] Cyberspace provides available resources for different information processing agents. Cognitive processes here need not be individual organism-centered, but rather human social practices-centered.

Artificial agents that can simulate human cognitive processes are outcomes of the extensions of human practices and the materialization of their logic space. The construction of the logical space allows humans to share resources, to communicate, and cooperate more reasonably, and to practice together as rational practitioners. Structured data generated by common practice can serve as resources in the

---

[143] Smart, "Extended Cognition and the Internet. A Review of Current Issues of Controversies," 362.

[144] As Clark stresses, "cognition is organism centered even when it is not organism bound." Andy Clark, *Supersizing the mind: Embodiment, Action and Cognitive Extension* (New York: Oxford University Press, 2008), 139.

[145] Ciano Aydin, "The artifactual mind: overcoming the 'inside–outside' dualism in the extended mind thesis and recognizing the technological dimension of cognition," *Phenomenology and the Cognitive Sciences* 14, no. 1 (2015):79.

[146] Ibid., 77.

[147] Ibid., 88.

[148] Smart, "Extended Cognition and the Internet. A Review of Current Issues of Controversies," 384.

environment to afford information processing agents the possibility to act in a reasonably interpretable way. The relationship between intelligent agents and the space in which their actions can be actualized is related to a deeper philosophical concern, namely the insight into the relationship between conceptual thought and experience in the German idealistic tradition, and in particular a Hegelian turn toward the mutual constitutive relation of rational agents and their reality.[149] This turn challenges the representationalist distinction between the intelligent mind as knower and the external reality as what to be known. Knowing agents and knowable realities are instead treated as outcomes of interaction in practice. The Hegelian turn thus involves an intersubjective and normative turn on the mind and knowledge. Intelligence as the capacity to enact procedural knowledge is placed here in a rationalist explanatory framework. In the next chapter, I will move into a philosophical exploration of how interpretable rational agents arise in shared practices. Particular attention is paid to how information patterns are formed in discursive practices to provide agents with the possibility of mutually interpretable rational action.

Bibliography

Aghaei S, Nematbakhsh MA, Farsan HK (2012) Evolution of the World Wide Web: from Web 1.0 to Web 4.0. Int J Web Semant Technol 3(1):1–10. https://doi.org/10.5121/ijwest.2012.3101

Aydin C (2015) The artifactual mind: overcoming the 'inside–outside' dualism in the extended mind thesis and recognizing the technological dimension of cognition. Phenomenol Cogn Sci 14(1):73–94

Aydin C, Woge MG, Verbeek P-P (2019) Technological environmentality: conceptualizing technology as a mediating milieu. Philos Technol 32:321–338. https://doi.org/10.1007/s13347-018-0309-3

Bailey KD (1994) Typologies and taxonomies. An introduction to classification techniques. Sage Publications, Thousand Oaks

Berners-Lee T (1998a) Relational databases on the Semantic Web. Design issues. Last modified 27 Aug 2009. https://www.w3.org/DesignIssues/RDB-RDF.html

Berners-Lee T (1998b) The World Wide Web: a very short personal history. http://www.w3.org/People/Berners-Lee/ShortHistory.html

Berners-Lee T (1998c) What the Semantic Web can represent. Design issues. Last modified 17 Sep 1998. https://www.w3.org/DesignIssues/RDFnot.html

Berners-Lee T (2006) Linked data. Design issues. Last modified 18 June 2009. https://www.w3.org/DesignIssues/LinkedData.html

Berners-Lee T, Fischetti M (2000) Weaving the Web: the original design and ultimate destiny of the World Wide Web by its inventor. Harper, New York

Brown, TB, Mann B, Ryder N, Subbiah M, Kaplan J, Dhariwal P, Neelakantan A, Shyam P, Sastry G, Askell A, Agarwal S, Herbert-Voss A, Krueger G, Henighan TJ, Child R, Ramesh A, Ziegler

[149] Rockmore points out that German idealism was rejected by early analytic philosophy because it had been misunderstood as denying the existence of a mind-independent external world. But after the criticism of empiricism in epistemology represented by Quine, Sellars, Putnam, Rorty, etc., analytic philosophy away from Hegel has shown a turn toward Hegel and German idealism. This turn is embedded in a broader pragmatic turn. Cf. Tom Rockmore, "Analytic Philosophy and the Hegelian Turn," *The Review of Metaphysics* 55, no.2 (2001): 342ff.

DM, Wu J, Winter C, Hesse C, Chen M, Sigler E, Litwin M, Gray S, Chess B, Clark J, Berner C, McCandlish S, Radford A, Sutskever I, Amodei D (2020) Language models are few-shot learners. arXiv:2005.14165. https://doi.org/10.48550/arXiv.2005.14165

Capurro R (1978) Information. Ein Beitrag zur etymologischen und ideengeschichtlichen Begründung des Informationsbegriffs. München, Saur Verlag

Capurro R, Hjørland B (2003) The concept of information. Annu Rev Inf Sci Technol 37:343–411

Clark A (2008) Supersizing the mind: embodiment, action and cognitive extension. Oxford University Press, New York

Clark A (2010) Memento's revenge: the extended mind, extended. In: Menary R (ed) The extended mind. MIT Press, Cambridge, pp 43–66

Clark A, Chalmers D (1998) The extended mind. Analysis 58(1):7–19. https://doi.org/10.1093/analys/58.1.7

Coppin B (2004) Artificial intelligence illuminated. Jones and Bartlett, London

Dafflon B, Moalla N, Ouzrout Y (2021) The challenges, approaches, and used techniques of CPS for manufacturing in Industry 4.0: a literature review. Int J Adv Manuf Technol 113:2395–2412. https://doi.org/10.1007/s00170-020-06572-4

Deckert R (2019) Digitalisierung und Industrie 4.0. Technologischer Wandel und individuelle Weiterentwicklung. Springer, Wiesbaden

Floridi L (2009a) Against digital ontology. Synthese 168:151–179. https://doi.org/10.1007/s11229-008-9334-6

Floridi L (2009b) Web 2.0 vs. the Semantic Web: a philosophical assessment. Episteme 6(1):25–37. https://doi.org/10.3366/E174236000800052X

Floridi L (2011) The philosophy of information. Oxford University Press, Oxford

Floridi L (2014) The 4th revolution: how the inforsphere is reshaping human reality. Oxford University Press, Oxford

Fodor JA (1983) The modularity of mind: an essay on faculty psychology. MIT Press, Cambridge

Gerber A, van der Merwe A, Barnard A (2008) A functional Semantic Web architecture. In: Bechhofer S et al (eds) The Semantic Web: research and applications 5021. Springer, Berlin, Heidelberg, pp 273–287

Gettier EL (1963) Is justified true belief knowledge? Analysis 23(6):121–123. https://doi.org/10.2307/3326922

Gibson JJ (1979) The ecological approach to visual perception. Houghton Mifflin, Boston

Graham M (2013) Geography/internet: ethereal alternate dimensions of cyberspace or grounded augmented realities? Geogr J 179(2):177–182. https://doi.org/10.1111/geoj.12009

Gruber TR (1993) A translation approach to portable ontology specifications. Knowl Acquis 5(2):199–220

Gruber TR (2009) Ontology. In: Liu L, Tamer Özsu M (eds) Encyclopedia of database systems. Springer, Boston, pp 1963–1965. https://doi.org/10.1007/978-0-387-39940-9_1318

Halpin H (2013) Social semantics. The search for meaning on the Web. Springer, New York

Haugeland J (1987) An overview of the frame problem. In: Pylyshyn ZW (ed) The robot's dilemma: the frame problem in artificial intelligence. Ablex Publishing Corporation, Norwood, pp 77–93

Hayes PJ (1980) The logic of frames. In: Frame conceptions and text understanding. DeGruyter, Berlin

Heath T, Bizer C (2011) Linked data: evolving the web into a global data space. Morgan & Claypool, San Rafael. HTML Version. http://linkeddatabook.com/editions/1.0/

Janlert L-E (1987) Modeling change: the frame problem. In: Pylyshyn ZW (ed) The robot's dilemma: the frame problem in artificial intelligence. Ablex Publishing Corporation, Norwood, pp 65–76

Kitchin R (2014) The data revolution: big data, open data, data infrastructures and their consequences. Sage, London

Lee J, Bagheri B, Kao H-A (2015) A cyber-physical systems architecture for industry 4.0-based manufacturing systems. Manuf Lett 3:18–23

Lehmann J, Isele R, Jakob M, Jentzsch A, Kontokostas D, Mendes PN, Hellmann S, Morsey M, van Kleef P, Auer S, Bizer C (2015) DBpedia—a large-scale, multilingual knowledge base extracted from Wikipedia. Semantic Web 6(2):167–195

Lopez TS, Ranasinghe DC, Harrison M, McFarlane D (2012) Adding sense to the Internet of Things: an architecture framework for smart objective systems. Pers Ubiquit Comput 16:291–308

Marcus G (2020) The next decade in AI: four steps towards robust artificial intelligence. arXiv:2002.06177. https://doi.org/10.48550/arXiv.2002.06177

McCarthy J, Hayes PJ (1969) Some philosophical problems from the standpoint of artificial intelligence. In: Meltzer B, Michie D (eds) Machine intelligence 4. Edinburgh University Press, Edinburgh, pp 463–502

Minsky M (1975) A framework for representing knowledge. In: Winston PH (ed) The psychology of computer vision. McGraw-Hill, New York, pp 211–227

Neuser W (2013) Wissen begreifen. Zur Selbstorganisation von Erfahrung, Handlung und Begriff. Springer VS, Wiesbaden

Norman DA (1993) Cognition in the head and in the world: an introduction to the special issue on situated action. Cogn Sci 17(1):1–6

Paulheim H (2017) Knowledge graph refinement: a survey of approaches and evaluation methods. Semantic Web 8:489–508

Peters JD (1988) Information: notes toward a critical history. J Commun Inq 12:10–24

Radford A, Jeffrey W, Child R, Luan D, Amodei D, Sutskever I (2019) Language models are unsupervised multitask learners. OpenAI Blog 1(8):9

Ray PP (2018) A survey on Internet of Things architectures. J King Saud Univ Comput Inf Sci 30(3):291–319. https://doi.org/10.1016/j.jksuci.2016.10.003

Roberts LG (1978) The evolution of packet switching. Proc IEEE 66(11):1307–1313

Rockmore T (2001) Analytic philosophy and the Hegelian turn. Rev Metaphys 55(2):339–370

Rowley J (2007) The wisdom hierarchy: representations of the DIKW hierarchy. J Inf Sci 33(2):163–180. https://doi.org/10.1177/0165551506070706

Rowley J, Hartley R (2016) Organizing knowledge: an introduction to managing access to information. Routledge, London

Russell B (1914) Our knowledge of the external world as a field of scientific method in philosophy. Open Court Publishing, Chicago

Smart PR (2012) The web-extended mind. Metaphilosophy 43(4):446–463

Smart PR (2017) Extended cognition and the internet. A review of current issues of controversies. Philos Technol 30:357–390

Strate L (1999) The varieties of cyberspace: problems in definition and delimitation. West J Commun 63(3):382–383. https://doi.org/10.1080/10570319909374648

Suchman LA (2007) Human-machine reconfigurations: plans and situated actions. Cambridge University Press, Cambridge

Sullivan D (2020) A reintroduction to our knowledge graph and knowledge panels. https://blog.google/products/search/about-knowledge-graph-and-knowledge-panels/

Szilagyi I, Wira P (2016) Ontologies and Semantic Web for the Internet of Things—a survey. In: 42nd annual conference of the IEEE industrial electronics society, pp 6949–54. https://doi.org/10.1109/IECON.2016.7793744

Thagard P (1984) Frames, knowledge, and inference. Synthese 61(2):233–259

The Linked Open Data Cloud (2022). https://lod-cloud.net/. Accessed 15 Jan 2022

Toepfer G (2011) Historisches Wörterbuch der Biologie. Geschichte und Theorie der biologischen Grundbegriffe. Band 2. Berlin: Springer

Verbeek P-P (2009) Ambient intelligence and the persuasive technology: the blurring boundaries between human and technology. Nanoethics 3:231–242

Wiegerling K (2024) Ubiquitous Computing und informatische Vernetzung. In: Gutmann M, Wiegerling K, Rathgeber B (eds) Handbuch Technikphilosophie. J.B. Metzler, Stuttgart, pp 345–354

William WA (1975) What's in a link: foundations for semantic networks. Bolt, Beranek and Newman, Cambridge

Williams D (2014) Models, metaphors and symbols for information and knowledge systems. J Entrep Manag Innov 10:79–107

Wittgenstein L (1953) Philosophical investigation. Translated by Anscombe GEM. Macmillan, New York

World Wide Web Consortium (W3C) (2004) Resource description framework (RDF): concept and abstract syntax. Last modified 10 Feb 2004. https://www.w3.org/TR/rdf-concepts/

World Wide Web Consortium (W3C) (2010) Linked open data. Last modified 7 June 2010. https://www.w3.org/egov/wiki/Linked_Open_Data

World Wide Web Consortium (W3C) (2012) Web ontology language. Last modified 11 December 2012. https://www.w3.org/2001/sw/wiki/OWL

World Wide Web Consortium (W3C) (2013) Semantic Web activity. Last modified 11 Dec 2013. https://www.w3.org/2001/sw/

World Wide Web Consortium (W3C) (2014a) RDF 1.1 prime. W3C working group note. https://www.w3.org/TR/rdf11-primer/

World Wide Web Consortium (W3C) (2014b) RDF 1.1 concepts and abstract syntax. https://www.w3.org/TR/rdf11-concepts/#section-triples

World Wide Web Consortium (W3C) (2016) W3C and big data. https://www.w3.org/blog/2016/10/w3c-and-big-data/

World Wide Web Consortium (W3C) (2021) W3C data activity building the web of data. https://www.w3.org/2013/data/. Accessed 7 June 2021

Zins C (2007) Conceptual approaches for defining data, information, and knowledge. J Am Soc Inf Sci Technol 58(4):479–493

Chapter 4
Triandic Interaction, Inter-Subjectivity, Inter-Objectivity

From Leibniz's idea of a universal language for presenting knowledge across disciplines to the current digitization and knowledge engineering supported by semantic technologies, the logical space of shared practices is materialized into the networked cyberspace that simulate and extends our possibilities for action. Semantic technologies are not only a class of ICTs, but also a simulation of knowing how to perform common practices. This chapter will discuss the conditions under which information that carries semantics can be formed and recognized in terms of formal semantics. This discussion will shed light on why interpretable information patterns and the shareable space they constitute are key to framing rational actions and their agents responsible for engaging in common discursive practices. It will begin with Davidson's modification of the Turing test, exploring the conditions required to confirm whether others can think. If semantics is to be taken as an important indicator of thinking, then a solid method is needed to figure out whether the expressions of others are meaningful and can be taken as true. Therefore, Davidson's truth-conditional semantics and theory of interpretation will be elaborated, which will lead to a discussion of the initial conditions for meaning that can be recognized and interpreted.

4.1 From the Imitation Game to a Play of Interaction

Recalling the key point of the Turing test, the question "Can machines think?" is strategically transformed into the question of whether machines can pass the imitation game. An important premise of the imitation game is that thinking can be determined through certain evidence. Such evidence must contain information that can convince the interrogator to make a judgment, rather than simply sensory data. Although the imitation game is often treated as a behaviorist method of grasping thought through observation, how the human interrogator makes judgments cannot

Y. Du, *Machines in Minds*, Studies in Applied Philosophy, Epistemology and Rational Ethics 78, https://doi.org/10.1007/978-3-032-15842-0_4

be fully reduced to a behaviorist approach, because the judgment of evidence depends on cognitive activities on the part of the human interrogator, such as understanding and interpreting evidence. To make reliable judgments and consistent interpretations, human interrogators need to constantly adjust their assumptions and attitudes toward what others express. How to provide human interrogators with a criterion for making reliable judgments through evidence is discussed by Davidson from an interpretivist standpoint.

In his article *Turing's Test*, Davidson proposes a methodological refinement of the original Turing test.[1] He begins by arguing that the Turing Test is in fact intended to test not only machines but also humans, i.e., to test "whether, and on what basis, normal human judges are able or willing to assign mental attributes to machines."[2] Accordingly, the focus of the test should shift from the test subject to the human interrogator. If there is a universal method and criterion for judging whether others are thinking based on their expressions, then this test can be applied to any object. Further, Davidson emphasizes that Turing's strategy is based on establishing observable evidence for determining thought, for "unless the presence of thought can be determined on the basis of external evidence, the question whether a machine can think has no special interest, since the same question will apply equally to everyone (else)."[3] That is to say, the Turing test represents a general strategy for determining whether other people can think by reference to external evidence. The advantage of this strategy is that the interrogator makes judgments on the basis of verbal expressions, without making prior assumptions about whether the test subject is a thinking subject.[4] The test subject can be judged to be a thinking subject only if its expression is recognized as meaningful by the interrogator. He thus agrees with the strategy that "Turing was right […] in taking as the only test for the presence of thought and meaning the interpretive powers and abilities of a human interpreter."[5] Human interpretive abilities become crucial to the test.

Nevertheless, Davidson notes that in the original version of the Turing test, the interrogator would encounter a significant issue, namely the lack of objective criteria for judgment. In the original version, the interrogator is isolated from the test subjects. This setting was intended to avoid other less important factors, such as appearance and voice, that might influence the interrogator's judgment. But separating the two parties from each other means that there is no opportunity to obtain any observable evidence from a shared environment. To determine whether the utterances of the test subjects are meaningful, the interrogator in the original version lacks actual evidence and relies on his or her own knowledge and psychological expectations. Davidson argues that purely subjective judgments are not reliable

[1] Donald Davidson, "Turing's Test [1990]," In *Problems of Rationality* (Oxford: Oxford University Press, 2004), 80f.

[2] Ibid., 80.

[3] Ibid.

[4] Ibid. As Davidson writes, "Turing's Test eliminates the possibility of telling *whether* a creature or machine thinks without determining *what* it thinks" (80).

[5] Ibid., 86.

because "[the interrogator] has no clue to the semantics of the object. There is no way he can determine the connection between the words that appear on the object's screen and events and things in the world."[6] From a semantic externalist view, Davidson suggests that figuring out meaning requires a shared environment where language use can be observed.

In consideration of this issue, Davidson modifies the original Turing Test. Unlike the original test in which a human interrogator is asked to choose between a human and a machine, the modified test involves only the interrogator and one test subject, both of whom are placed in a shared public environment. The two sides of the conversation and the environment they share can form a three-way interaction, which is between the test subject and the environment, between the test subject and the interrogator, and between the interrogator and the environment. The triadic interaction ensures that the semantics of the test subject can be interpreted by the interrogator in a reasoned manner. Because Davidson holds that the criterion for determining whether a test subject thinks should be the semantics of it's expressions, and semantics deals with truth. So semantic evidence should be able to convince the interrogator that the test subject can use language to establish referential relations and that its expression can be true under certain conditions. He writes:

> What is needed is evidence that the object uses its words to refer to things in the world, that its predicates are true of things in the world, that it knows the truth conditions of its sentences. Evidence for this can come only from further knowledge of the nature of the object, knowledge of how some of its verbal responses are keyed to events in and aspects of the world, events and aspects also known to the interrogator.[7]

Semantics is understood here as a referential relation that is directed to objects or events in the world. An expression possessing semantic properties carries this relation, and it is true when it satisfies the corresponding conditions. For one individual to understand the referential relation of the other, they need to be in the same environment. Thus, for the interrogator to perceive and interpret the semantics of the test subject, the interrogator should be allowed to engage in the interaction with the tested subject in a shared public world. This is the reason why the Turing test is modified to be set in a long-term triadic interaction. The triadic interaction is the condition for language use to be meaningful in an observable way and interpretable in terms of causal relations. As he explains:

> For the object to have a semantics, it must operate in the world in a certain way, and for someone else to grasp those semantics, there must be a three-way interaction among object, interrogator and a shared world. [...] The object must be brought into the open so that its causal connections with the rest of the world as well as with the interrogator can be observed by the interrogator.[8]

Sustained triadic interactions facilitate the human interrogator to participate, observe, and interpret the construction of referential relationships. By taking part in

---

[6] Ibid., 82.

[7] Ibid., 83.

[8] Ibid., 83f.

the interaction, the interrogator can interpret the causal connection between the interactor's actions and the environment in terms of physical relations on the one hand, and the referential relations, attitudes, and intentions of the speaker's language use in terms of mental vocabulary on the other. These two interpretive schemes complement but are not reducible to each other, while both are committed to the existence of the objects they explain. This may appear to be a dualistic treatment of the mind–body problem, but Davidson holds an anomalous monistic view in which it is argued that "each object or event identified in mental terms is identical with an object or event identifiable in physical terms, though the classes staked out by the mental vocabulary cannot be equated by definition or strict laws with classes definable in the physical vocabulary."[9] According to this view, the inquiry into the mind–body problem turns into a question of the legitimate scope for the use of physical and mental terms. To understand the commitments that different terms take on when they are used and translated into each other, it is necessary to gain a deep understanding of his theory of interpretation.

The modified Turing test turns the test into a long-term triadic interaction between the interrogator, the test subject, and their shared world. The idea behind this modification is that if an expression can be interpreted as meaningful, then it can be jointly taken as true by the speaker and the interpreter. And expressions that can be recognized and interpreted as true are generated in triadic interactions between multiple language users and a shared world. The key to understanding the intersubjective roots of meaning-making and giving is Davidson's theory of meaning and interpretation, which is the focus of the next subchapter.

4.2 Recursive Patterns of Meaning-Truth-Beliefs for Interpretation

What makes an expression meaningful? This question will be explored in this chapter in three steps. To begin with I will elaborate Davidson's theory of meaning, i.e., truth-conditional semantics. According to this theory of meaning, the question of whether an expression has meaning is converted into the question of whether this expression can be true under certain conditions. This shift turns the linguistic representation of meaning into an interpretation of the truth conditions at the metalinguistic level. Next, I will clarify the available patterns that are necessarily presupposed in interpretation through Quine's thought experiment of radical

[9] Donald Davidson, "Indeterminism and Antirealism [1997]," in *Subjective, Intersubjective, Objective* (New York: Oxford University Press, 2001), 72. The anomalous monism was proposed to provide a coherent explanation for the interaction between mental and physical events. Anomalous monism assumes that the token of every mental event can be the same as the token of some physical event. But the types of mental events are different from the types of physical events. Also see, Donald Davidson, "Mental Events [1970]," In *Essays on Actions and Events: Philosophical Essays Volume 1* (Oxford: Oxford University Press, 2001), 214.

translation and related debates. Patterns that are committed to be interpretable afford beliefs of something that can be taken to be true. At last, I will introduce the strategy for interpretation suggested by Davidson, the principle of charity, which treats the interpreted expression as potentially true and the one who utters it as a rational speaker. Interpretable patterns that carry meaning, truth, and belief are commonly committed to realities in intersubjective language use. The attitude of commitment to their interpretability is the starting point for the principle of charity.

4.2.1 Interpreting Meaning with Truth-Conditional Semantics

Literally, meaning is what signs express. In a language, some combinations of characters make sense and others do not. Skilled language users of a language can easily distinguish them. How are meanings presented by expressions, and how are they interpreted and grasped by language users? To account for these issues, Davidson attempts to explain meaning in the way that formal semantics studies the available linguistic structures that carry semantics. A widely accepted semantic holism suggests that the meaning of a word or sentence cannot be understood in isolation, but only in relation to other parts of the language, or even to the whole language.[10] However, if the meaning of an expression depends entirely on its relation to other expressions in the language, then one would have to learn every sentence in the language to figure out the meaning of one of them.[11] In fact, language users begin to master a language by learning individual expressions, and they only need to learn some of the expressions in that language in order to acquire the meaning of the others. This indicates the compositional nature of meaningful linguistic expressions, which allows that by learning only a limited number of expressions, one can master a language with theoretically infinite expressions. Hence, an adequate theory of meaning must account for its compositionality. This is the goal of Davidson's theory of meaning, not to specify what meaning is, but to describe how meaning is given.

The first step in constructing a theory of meaning is to find the basic structure used for composition. Davidson holds that the compositionality of meaning shows that language use must have some basic recursive structure which, once mastered, can be used to represent and interpret the meaning of various expressions in that language.[12] He begins by assuming that any presentation of meaning in a language

---

[10] Donald, Davidson, "Semantics for Natural Languages [1970]," In *Inquiries into Truth and Interpretation* (Oxford: Oxford University Press, 2001), 64.

[11] Donald, Davidson, "Truth and Meaning [1967]," In *Inquiries into truth and interpretation* (Oxford: Oxford University Press, 2001), 22. He notes: "If sentences depend for their meaning on their structure, and we understand the meaning of each item in the structure only as an abstraction from the totality of sentences in which it features, then we can give the meaning of any sentence (or word) only by giving the meaning of every sentence (and word) in the language." (22)

[12] Ibid., 23. Recursive structures are understood here literally as structures that can be repeatedly applied to form other structures.

L can be generalized to "*s* means *m*", where *s* is a sentence in L and *m* is its meaning.[13] The general sentence structure connected by the predicate "means" is such a recursive structure that it is possible to form "*s* means *m*", "'*s* means *m*' means *m*", and so forth. A theory of meaning for a language "must entail *all* sentences of the form '*s* means *m*'," but he then notes that the term "means…" is often used in an ambiguous way, as it can either refer to extensions or signify intensions.[14] Broadly speaking, extension defines the meaning of a word by listing what is referred to, while intension defines the meaning of a word by describing its properties and conditions. To avoid the infinite regress that a purely connotative definition of meaning might lead to, he reformulates the general structure as "*s* means *p*", where *p* is extensional and can be any sentence in the metalanguage that the interpreter uses to interpret meaning. A theory of meaning for a language L then becomes a matter of finding a way to assign a corresponding translation *p* in the metalanguage to each sentence *s*. He writes:

> The theory will have done its work if it provides, for every sentence *s* in the language under study, a matching sentence (to replace '*p*') that, in some way yet to be made clear, 'gives the meaning' of *s*. One obvious candidate for matching sentence is just *s* itself, if the object language is contained in the metalanguage; otherwise a translation of *s* in the metalanguage. As a final bold step, let us try treating the position occupied by '*p*' extensionally: to implement this, sweep away the obscure 'means that', provide the sentence that replaces '*p*' with a proper sentential connective, and supply the description that replaces '*s*' with its own predicate. The plausible result is [(T)]. The plausible result is

$$(T) \quad s \, \text{is} \, T \text{ if and only if } p.$$

What we require of a theory of meaning for a language L is that without appeal to any (further) semantical notions it places enough restrictions on the predicate 'is T' to entail all sentences got from schema T when 's' is replaced by a structural description of a sentence of L and 'p' by that sentence.[15]

The strategy is to replace "*s* means *p*" with the structure (T) "*s* is *T* if and only if *p*", converting "means" into a meta-linguistic description of the biconditional relation. Structure (T) is adapted from the T-sentence "*s* is true if and only if *p*" from Tarski's semantic theory of truth. The advantage of this strategy is that *p* as the condition for satisfying *s* to be true is a general description of their relation, allowing *p* not to be reduced to any specific object. In this way, the theory of meaning can universally describe the assignment of meaning to any sentence in a language.

Davidson's truth-conditional theory of meaning is in fact an inversion of Tarski's semantic theory of truth. While Tarski provides a formal definition of the concept of truth in terms of meaning, Davidson provides a formal definition of meaning in terms of truth. Tarski defines the concept of truth for formal languages in order to

---

[13] Donald Davidson, "Theories of Meaning and Learnable Languages [1965]," In *Inquiries into Truth and Interpretation* (Oxford: Oxford University Press, 2001), 8f.

[14] Davidson, "Truth and Meaning," 23.

[15] Ibid.

avoid logical contradictions. On the one hand, he proposes an undefinability theorem, which states that the semantic truth of a formal system cannot be defined by the system itself.[16] On the other hand, he distinguishes metalanguage from object language, in which the object language is the language being expressed, and the metalanguage is the language used to talk about the object language. Accordingly, it is reasonable to assume that a metalanguage that can describe the object language includes a wider range of statements and explanatory power than the object language. By defining the semantic truth of the object language in terms of the metalanguage, the semantic contradictions arising from self-referential linguistic structure can be avoided. Thus, semantic truth can be represented in a formal relation, i.e., a T-sentence consisting of an object language and a metalanguage. A T-sentence can be written as, for example, "Schnee ist weiss" is true if and only if snow is white, where "Schnee ist weiss" is the object language and snow is white is its translation into the metalanguage. By means of the T-sentence structure, Tarski proposes to define truth predicates in a formal language by satisfying Convention-T.[17] A theory of truth satisfies Convention-T when every instance of "*s* is true if and only if *p*" can be consistently derived within it. That is to say, the truth predicate in a formal language can be materially adequately defined as long as any true sentence in that language can be assigned an instance, i.e. the condition that the sentence is true can be expressed in metalanguage.[18]

In Davidson's approach to defining meaning, by contrast, the convention to be satisfied is that any meaningful sentence in a language can be translated at the metalinguistic level as true under certain conditions. The grasp of truth is here the starting point for understanding meaning. As we have seen, in Tarski's approach to defining the concept of truth, one does not need to inquire why the sentence "Schnee ist weiss" means, or can be translated as, snow is white. This is because, influenced by logical positivism, the metalanguage used for understanding is supposed to contain the common sense understanding of meaning. But this is precisely the key question in Davidson's theory of meaning, why the sentence "Schnee ist weiss" can be given such a meaning, or how it can be interpreted to mean that. One reason for this reversal is that Tarski aims to define the semantics of truth predicates for formal languages, whereas Davidson aims to provide a descriptive theory of semantics for natural languages based on the concept of truth.[19] Taking truth as a basic concept for

---

[16] Tarski, "The Concept of Truth in Formalized Languages," 152.

[17] Ibid., 187.

[18] Ibid. Tarski distinguishes between formal adequacy conditions and material adequacy conditions for a theory of truth, the former is the internal consistency of the theory, and the latter is satisfying Convention-T.

[19] Davidson later explains the motivation for this reversal: "I considered truth to be the central primitive concept, and hoped, by detailing truth's structure, to get at meaning. These are remarks about theories of truth, of course, not remarks to be found in them. Something else that was slow coming to me was that since I was treating theories of truth as empirical theories, the axioms and theorems had to be viewed as laws. So a theorem like "'Schnee ist weiss' is true in the mouth of a German speaker if and only if snow is white" has to be taken not merely as true, but as capable of supporting counterfactual claims." Davidson, *Inquiries into truth and interpretation*, xvi.

understanding meaning is taking meaning to be externalized as empirically observable and testable. Therefore, Davidson's core strategy is to convert the meaning of a sentence into conditions for it to be true as explained from the meta-linguistic level.

The intertwining relationship between meaning and truth has become obvious. In Davidson's theory of meaning, Convention-T is used to provide "a clear and testable criterion of an adequate semantics for a natural language".[20] This approach does not directly define meaning, but rather examines whether a sentence makes sense by interpreting the conditions under which it is true at the meta-linguistic level. By constructing a T-sentence, a sentence in the object language is allowed to be interpreted by the metalanguage. Interpreting the object language from the truth-conditional relation ensures the consistency of the meta-linguistic interpretation. Accordingly, the semantics of a language is not grasped within that language itself, as semantic holism holds, but at the meta-linguistic level. Returning to the question of how one acquires a language by learning only part of it, the answer is by learning part of it one can acquire the ability to interpret the truth conditions of the object language from the meta-linguistic level. T-sentences composed of meta- and object languages provide the basic recursive structure for the compositionality of language. Therefore, the meaning of an expression cannot be simply reduced to the referential relations between words and entities, but is in an interpretive relation between the meta- and object language.[21]

4.2.2 Indeterminate Translation and Radical Interpretation

According to truth conditional theory of meaning, the question of whether a sentence has meaning is converted into the question of whether that sentence can be true under certain conditions, which need to be interpreted at the meta-linguistic level. But what exactly is the meta-linguistic level? By definition, a metalanguage is a language used to interpret the object language.[22] Using a metalanguage to interpret usually implies that the metalanguage is already understandable, while object language needs to be translated. Yet the theory of meaning is concerned with how true conditions are translated into metal language in a way that is understandable. For an unheard expression like "Gavagai", constructing a T-sentence "'Gavagai' is true if and only if…", how can one determine the conditions, if any, under which it is true? This example is given by Quine in his thought experiment on radical translation.

[20] Davidson, "Truth and Meaning," 35.

[21] Donald Davidson, "In Defence of Convention T [1973]," In *Inquiries into truth and interpretation* (Oxford: Oxford University Press, 2001), 66.

[22] Commonly, a metalanguage can be a mother tongue while the object semantics is a foreign language, since it is assumed that one's first language has a richer explanatory vocabulary than a later acquired language. But more generally, a metalanguage is the language used for understanding while the object language is the language being interpreted. In this sense, they can be expressions of the same language that have been mastered and those that have not yet been acquired.

Suppose a field linguist is studying the language of an unfamiliar tribe. A rabbit runs past and at the same time a native utters the expression "Gavagai". Without any previous knowledge of the native language, how would the linguist translate this expression? This difficulty leads to Quine's thesis on the indeterminacy of radical translation, and Davidson's account for radical interpretation.

I begin by illustrating Quine's thesis on radical translation, where "observation sentences" are taken as the starting point of translation. The concept of observation sentences is introduced by Quine to refer to sentences that are caused by physical stimuli and do not require reasoning, such as "Cold!", "Rabbit!", etc.[23] Quine believes that observation sentences are the entrance for children to learn language, and that translating an unknown language begins with learning its observation sentences.[24] Returning to the thought experiment of translation, both "Rabbit" and "Gavagai" can be thus treated as observation sentences in their respective languages:

> The observation sentence 'Rabbit' has its stimulus meaning for the linguist and 'Gavagai' has its for the native, but the affinity of the two sentences is to be sought in the externals of communication. The linguist notes the native's utterance of 'Gavagai' where he, in the native's position, might have said 'Rabbit'. So he tries bandying 'Gavagai' on occasions that would have prompted 'Rabbit' and looks to native for approval. Encouraged, he tentatively adopts 'Rabbit' as translation.[25]

Yet translation from observation sentences would give rise to the difficulty of indeterminate reference. It is conceivable that the observation sentences in this unfamiliar language refer to something quite different from what the linguist's own language refers to. "Gavagai" might not express the whole rabbit, but "undetached rabbit parts", "rabbit stages", etc.[26] The reason for this is that although observation sentences are assumed to be directly caused by physical stimuli, they are a report rather than a direct response to stimuli.[27] Sentences as reports do not describe atomic facts in isolation, but compose a context in which their meanings are interdependent. Even if an observation sentence independently refers to evidence, its meaning needs to be confirmed in the set of sentences it consists of, rather than reduced to physical stimuli. Thus, on the same physical stimuli occasion, there can be different

---

[23] An observation sentence is a component of language rather than a grammatical structure, so it can be either a word or a complete sentence. Even if an observational sentence has only one word, it is also expressing a sentence, or more precisely, an assertion, such as "'Rain', 'Cold', or 'Rabbit', for 'It's raining', 'It's cold', 'It's a rabbit'." Cf. Willard V.O. Quine, *Pursuit of Truth* (Cambridge, MA: Harvard University Press, 1992), 3f.

[24] Observation sentences are thus treated as the linguistic means of acquiring knowledge. He writes, "[o]bservation sentences are the bottom edge of language, where it touches experiences: where speech is conditioned to stimulation. It is ultimately through them that language in general gains its meaning, its bearing on reality." Willard V.O. Quine and Joseph S. Ullian, *The Web of Belief* (New York: McGraw-Hill, 1978), 28.

[25] Quine, *Pursuit of Truth*, 42.

[26] Willard V. O. Quine, *Ontological Relativity and Other Essays* (New York: Columbia University Press, 1969), 28.

[27] Quine, *Pursuit of Truth*, 61. The concept of observation sentences is an inheritance and adaptation of Carnap's concept of protocol sentences that report the sensations of an observer.

observation sentences. “Gavagai” can refer to either a part or a state of the rabbit, which makes no observable difference to the linguist. In the case of radical translation, without any knowledge about the target language, it is impossible for the linguist to confirm which translation is more correct. This difficulty is a basic case of translation indeterminacy, also known as indeterminacy of reference.

The indeterminacy of reference does not mean that translation is impossible. Quine argues that hypotheses about observation sentences can eventually be confirmed by the native speaker’s approval or disapproval, even if the reference of each sentence cannot be precisely identified.[28] The linguist can compile a collection of observation sentences into a translation manual through the speaker’s approval and disapproval behaviors. A translation manual is a hypothesis about what observation sentences in an unknown language mean. Depending on different hypotheses, different translation manuals can be developed. In this way, there can be different translation manuals that are incompatible with each other, but respectively internally consistent and approved. Quine writes:

> The thesis is then this: manuals for translating one language into another can be set up in divergent ways, all compatible with the totality of speech dispositions, yet incompatible with one another. In countless places they will diverge in giving, as their respective translations of a sentence of the one language, sentences of the other language which stand to each other in no plausible sort of equivalence however loose. The firmer the direct links of a sentence with non-verbal stimulation, of course, the less drastically its translations can diverge from one another from manual to manual.[29]

There can be significant divergence between different manuals as hypotheses to guide translation. It is difficult to decide which manual is better because the reference of the underlying observation sentences is indeterminate. This new problem is a stronger form of translation indeterminacy, and Quine calls it holophrastic indeterminacy.[30] Yet even though no single manual can be determined to be completely correct, the linguist can still use them to guide communication with native speakers. The thesis of translation indeterminacy in general does not imply that there are no correct translations. Quine argues that indeterminacy reveals exactly that it is the expressions that can be translated in different ways that are meaningful, because he rejects reducing the meaning of an expression to a fixed substantial reference but considers meaning to be a theoretical object that is presented by language. In order to translate “Gavagai” into a meaningful expression, the linguist needs to assume that it does make sense and employs “‘Gavagai’ is true if and only if…” to construct hypotheses about “Gavagai” and different translation manuals. In other words, as the committed theoretical object, meaning is assumed here to be presented in different hypotheses.

---

[28] It is questionable here how the linguist can observe and confirm others’ expressions about approval and disapproval without relevant knowledge. Quine argues that no background knowledge is needed here, and that empathy is the basic ability to learn about others’ expressions of approval and disapproval. Ibid., 42f.

[29] Willard V. O. Quine, *Word and Object* (New York: MIT Press/Wiley, 1960), 27.

[30] Quine, *Pursuit of Truth*, 50.

However, Davidson notes that using translation manuals to theorize observation sentences in the unknown language can lead to difficulties in comprehension. He points out that the translation approach with manuals involves the linguist's mother tongue, the object language being translated, and a third language, i.e., a metalanguage used for theorizing.[31] In other words, the hypothesis about observation sentences is distinguished from expressions that the linguist has already mastered, and the manual is written in a theoretical metalanguage rather than a natural language. In the translation process, it is assumed that the linguist theorizes the object language in the metalanguage and compiles it into a manual, then in order to understand this manual, the metalanguage needs to be translated again into the linguist's own language. This may lead to a case in which the manual cannot be translated fluently, or even intelligibly, into the linguist's own language. That is, the linguist may not fully understand the theorized observation sentences in the metalanguage, because of the difference between the metalanguage and her own language. Thus, translation manuals consisting of theorized observation sentences may not really help the linguist to understand the object language.

This rebuttal suggests that the metalanguage used for translation should be a language for comprehension, not a purely theoretical construction. This brings us back to the question of what it means to be at the meta-linguistic level. For Quine, the meta-linguistic level is the theoretical level about meaning, and metalanguages are theories. For Davidson, the meta-linguistic level is the level of understanding and giving meaning, and the metalanguage is the language used for comprehension and interpretation. A radical translation must be a radical interpretation for the latter. In the usual case, there is no need to theorize about unfamiliar expressions as in radical translation. People usually interpret unfamiliar expressions in terms of expressions they have already mastered. Accordingly, in the case of radical translation, the linguist can already interpret unknown expressions in some way before having a well-developed translation manual. At least expressions that indicate approval and disapproval are supposed to be interpreted in the process of communication with native speakers. Unlike Quine's aim of revealing that different languages can be translatable but indeterminately different conceptual systems, Davidson is concerned with the formation of intelligible meaning between subjects, regardless of whether they use different conceptual systems.

This brings us back to Davidson's theory of meaning, in which meaning can converted into truth conditions that are interpreted in a metalanguage. To interpret an unknown expression is to figure out whether the expression has meaning and

[31] He writes: "In the general case, a theory of translation involves three languages: the object language, the subject language, and the metalanguage (the languages from and into which translation proceeds, and the language of the theory, which says what expressions of the subject language translate which expressions of the object language). And in this general case, we can know which sentences of the subject language translate which sentences of the object language without knowing what any of the sentences of either language mean (in any sense, anyway, that would let someone who understood the theory interpret sentences of the object language)." Donald Davidson, "Radical Interpretation [1973]," in *Inquiries into truth and interpretation* (Oxford: Oxford University Press, 2001), 129.

what meaning it has, i.e., figuring out under what conditions, if any, the expression can be true. Interpretation does not require theorizing about unknown observation sentences, but rather ascribing meaning to the object language in the interpreter's own words.[32] Whatever "Gavagai" actually means, it only makes sense to the linguist if she first treats it as being within her own recognition. The focus of interpretation is not to compare differences between the referential systems of two languages, but to present the meaning of the object language in the known language. One can form T-sentences of the object language as a tool for interpretation, e.g., "'Gavagai' is true if and only if a rabbit appears", "'Gavagai' is true if and only if something runs past", etc.[33] The key to forming a T-sentence is to put unknown utterances like "Gavagai" into the sentence structure "'*s* is true if and only if *p*", that is, assuming that the unknown expression can be taken as true under certain conditions. As discussed in the previous section, Davidson's theory of meaning is a reversal of Tarski's theory of truth. Here, Davidson explains the advantages of using truth as the basic concept for interpreting unknown expressions:

> What I propose is to reverse the direction of explanation: assuming translation, Tarski was able to define truth; the present idea is to take truth as basic and to extract an account of translation or interpretation. The advantages, from the point of view of radical interpretation, are obvious. Truth is a single property which attaches, or fails to attach, to utterances, while each utterance has its own interpretation; and truth is more apt to connect with fairly simple attitudes of speakers.[34]

Truth is understood here in two ways as the basis for interpretation. On the one hand, truth refers to the underlying recursive structure of languages in the form of T-sentences. The recursive structure of T-sentences undertakes the commitment that an expression can be interpreted as true at the meta-linguistic level. On the other hand, the concept of truth is directly linked to a speaker's attitude of taking the expression as true. As Davidson writes, "a speaker holds a sentence to be true because of what the sentence (in his language) means, and because of what he believes."[35] That an expression can be interpreted by a metalanguage as true under certain conditions, and the speaker's beliefs about it, constitute two aspects of the concept of truth. For an interpreter to figure out what an expression means, it is necessary to first hold the attitude that this expression is meaningful, i.e., that it can be true under certain conditions. On this basis the interpreter can construct the T-sentence of the expression. Thus, holding the attitude that an expression can be true is the starting point for radical interpretation:

> A good place to begin is with the attitude of holding a sentence true, of accepting it as true. This is, of course, a belief, but it is a single attitude applicable to all sentences, and so does not ask us to be able to make finely discriminated distinctions among beliefs. It is an attitude an interpreter may plausibly be taken to be able to identify before he can interpret,

[32] Davidson, "Radical Interpretation," 130.

[33] Ibid.

[34] Ibid., 134.

[35] Ibid.

> since he may know that a person intends to express a truth in uttering a sentence without having any idea *what* truth.[36]

Holding the attitude of taking it as true is a primitive state, no matter what is specifically expressed. Treating an expression as if it can be true allows it to be the basic pattern that can be interpreted as meaningful. In the case of "Gavagai", the linguist first needs to treat "Gavagai" as a meaningful expression to construct "'Gavagai' is true if and only if *p*". Meanwhile, since the meaning of the expression and the speaker's attitude are interdependent through the concept of truth, the linguist must also assume that the speaker holds such an initial attitude when uttering "Gavagai". This interpretive strategy requires the interpreter not only to account for the corresponding evidence, but also to give a logically consistent account of the speaker's beliefs. The next section will elaborate on the implementation of this interpretive strategy and show why it ensures rational understanding and communication.

4.2.3 *Principle of Charity and Commitment to an Inter-Subjective Reality*

As we have seen, the initial attitude of taking an expression as true makes it possible to interpret an expression whose reference is indeterminate. The problem of radical translation is transformed into radical interpretation in this way. The interpreter needs to interpret whether an expression can be true under certain conditions, namely, to both interpret the meaning of the expression and attribute the beliefs of the speaker. Based on the concept of truth, the meaning of an expression and the speaker's beliefs about it are two interdependent variables. In this regard, Davidson proposes the principle of charity as an explanatory strategy, that is, always assuming either that the speaker has a true belief about the expression, or that the content of the expression is determinate.

In general, implementing the principle of charity means that interpretation begins by adopting the attitude that the expression can be true under certain conditions and that the speaker has semantics. Simply put, it is assumed that the speaker uses language in the same rational manner as the interpreter. To facilitate beginning interpretation, the interpreter should attribute the speaker's beliefs to be as true as possible and interpret the expressions as meaningful as possible.[37] Specifically, the principle of charity involves coherence and correspondence. The principle of coherence requires the interpreter to endow expressions of the speaker with logical consistency in expression, while the principle of correspondence allows the interpreter

[36] Ibid., 135.

[37] Donald Davidson, "On the Very Idea of a Conceptual Scheme [1974]," In *Inquiries into Truth and Interpretation* (Oxford: Oxford University Press, 2001), 197.

to hold that the speaker can have true beliefs about the world.[38] Combining these two aspects, the speaker can be treated as someone with basic rationality that can be understood by the interpreter. This is the crucial point of the principle of charity as an interpretative strategy, treating the unknown as interpretable and extending subjective understanding to intersubjective interpretation. In particular, the principle of charity states that meaning understanding is interpretation between subjects rather than individual representations of the outside world. According to this interpretive strategy, the recursive pattern that carries the meaning, truth, and beliefs open to interpretation is the available pattern for generating meaning. It can be understood in the following aspects.

First, meaning that is converted into truth conditions cannot be reduced to physical stimuli or an individual's reception of it. That is to say, the assignment and interpretation of meaning cannot be reduced to cause and effects. Interpretation begins with taking the attitude of treating an expression as if it can be meaningful. This is a normative constraint on intelligibility. Thus, the principle of charity does not require the interpreter to assume that the speaker has a similar material basis for receiving physical stimuli in be intelligible, but rather that the way the speaker uses language is rational, namely, coherent and correspondent. Second, a theory of meaning based on the concept of truth is a pragmatic account of the triad of meaning, truth, and belief.[39] Neither the correspondence theory of truth alone nor the coherence theory of truth alone is sufficient to account for how a speaker consistently holds correct beliefs about the world. While correspondence theory holds that the truth of an expression is determined by whether it corresponds to fact, coherence theory holds that the truth of an expression is determined by the whole language. Bringing the two together, the concept of truth can serve as an empirical criterion that regards the consistency of the speaker's beliefs and their conformity to facts. Third, constructing T-sentences to satisfy Convention-T is the way to implement the principle of charity with this concept of truth. To clarify the truth conditions of one expression, the truth conditions of other expressions related to it need to be clarified to satisfy Convention-T. Interpretation is an ongoing process of constructing consistent sets of T-sentences, which are not hypotheses to guide translation, but coherent understandings expressed in the language of the interpreter.

However, radical interpretation does not avoid the problem of indeterminacy. For the interpreter to understand coherently, the interpretation may differ from what exactly the speaker meant. The speaker can still be misunderstood, because the interpreter implementing the principle of charity is centered on her own understanding and language use. The problem of indeterminacy here touches on a more fundamental issue than the understanding of meaning, that is, the necessity of converting meaning into intersubjectively recognizable truth conditions. The translation

[38] Donald Davidson, "Three Varieties of Knowledge [1991]," In *Subjective, Intersubjective, Objective* (New York: Oxford University Press, 2001), 211.

[39] Rorty also sees Davidson's work as an approach to move away from the representationalism. Cf. Richard Rorty, "Pragmatism, Davidson and Truth," In *Truth and Interpretation. Perspectives on the philosophy of Donald Davidson*, ed. Ernest Lepore (Cambridge: Blackwell, 1986), 348ff.

indeterminacy thesis deals not only with how something is presented in different languages, but with whether there is indeed something objective that is independent of its different representations.

In fact, Quine attempts to shed light on the issues of ontological relativity and ontological commitment between different languages through the thesis of the indeterminacy of translation. In the case of radical translation, we have seen that the indeterminacy of translation results from failing to determine the exact reference of an observation sentence. The indeterminacy of reference leads to the indeterminacy of a holophrastic translation. There can be different translation manuals, each of which is internally consistent and can be confirmed by behavioral evidence, so that the translator cannot determine which one is more correct. The indeterminacy of reference as the underlying problem is also called ontological relativity, because this problem represents the fact that "what there is" cannot be determined in terms of "semantic ascent". According to Quine, "semantic ascent" turns ontological questions into linguistic objects that can be talked about by words and sentences.[40] An ontological question is about "what there is", by contrast, a semantic ascent ontological question is about what presuppositions are implicit in the expression "what there is". He argues that talking about "what there is" in different languages can be translated by existential quantification in first-order logic.[41] Accordingly, "(there is a) rabbit" can be translated as "there is something x, and x is a rabbit". In this way, the existence of something x ($\exists x$) becomes a variable, in his words, "to be is to be the value of a variable. More precisely, what one takes there to be are what one admits as values of one's bound variables."[42]

The advantage of the semantic ascent strategy is to ensure that the ontological presuppositions of different languages, although relative to each other, all contain ontological commitments to some expressible being. In translating different languages or comparing different theories, the ontological commitments assumed can be revealed. In radical translation, there is no way to determine the exact reference of "Gavagai" because there may be different ways to theorize the stimuli, but it is necessary to assume that there is indeed something that causes the stimuli. The translator first assumes that the native speaker receives similar stimuli from a running rabbit, and then theorizes about it and compiles manuals accordingly, where "rabbit", "part of a rabbit" or "stage of a rabbit" are different values for the existence of something x.[43] The theorization of observation sentences can be written as first-order logic containing existential quantifiers, which always undertake

---

[40] Quine, *Pursuit of Truth*, 81.

[41] He explains: "It has been objected that what there is a question of fact and not of language. True enough. Saying or implying what there is, however, is a matter of language; and this is the place of the bound variables. [...] It has been objected that the logical notation of quantification is an arbitrary and parochial standard to adopt for ontological commitment. The answer is that the standard is transferable to any alternative language, insofar as we are agreed on how to translate quantification into it." Ibid., 27.

[42] Ibid., 26.

[43] Ibid., 31.

commitments to the variable that there is something *x* in a semantic ascent sense. In turn, talking about ontological relativity only makes sense if it is committed to something really existing for different ways of quantification.[44] It also implies that translating a language not only requires theorizing its basic presuppositions at the meta-linguistic level, but also acknowledging that the basic presuppositions of one's own language are not unique. Even if the underlying references of both languages cannot be translated one-to-one, it is necessary to admit that both express commitments to something that can be recognized.

Ontological commitments are significant for information communication between different conceptual networks. As introduced at the end of Sect. 3.1, information ontologies play important roles in knowledge representations, because they as a set of explicit terminological prescriptions provide conceptual relations for concrete assertions in knowledge bases. Different knowledge representations can have different ways of conceptualization, and a term referring to the same fact may be prescribed variably in different knowledge bases. Agreements are needed for consistency in the communication of shared vocabularies in different knowledge representations. These agreements are ontological commitments that ensure coherent and correspondent translation in cross-domain communication. Thus, ontological commitments in knowledge representations are agreements and standardizations of conceptualizations specified at the meta-linguistic level across domains. Here, what is committed is what can be represented by different conceptual networks, as Gruber states "what 'exists' is precisely that which can be represented".[45] With the assurance of ontological commitments, different knowledge representations mutually recognize each other's conceptualization as communicable and translatable. Recalling the discussion of the relationship between information and reality in Sect. 3.1, information can be understood three-fold in terms of being about reality, for reality, and as reality itself. Communicable information processing commits to a reality that can be presented by structured information that exists independently of it. This is what Floridi calls minimal ontological commitments.[46] In line with Quine's approach to ontological issues, the issue of what there is can be dealt with as a commitment to a reality that can be presented by logical structures. Accordingly, knowledge bases that can engage in information exchange necessarily recognize each other's ontological commitments to objects in their respective domains. Different knowledge representations mutually acknowledge each other's ways of conceptualization in the communication and allow the objects they present to be

[44] Quine even argues that ontology only makes sense when it is talked about in terms of different theories, and between translations in different languages. "Ontology is thus meaningless for a theory whose only quantification is substitutionally construed; meaningless, that is, insofar as the theory is considered in and of itself. [...] The answer depends on both theories and, again, on the chosen way of translating the one into the other." Quine, *Ontological Relativity and other Essays*, 64.

[45] Gruber, "A Translation Approach to Portable Ontology Specifications, 200.

[46] Luciano Floridi, *The Philosophy of Information*, 340.

translated in a universe of discourse they share with other knowledge representations.

It is worth stressing again that ontological relativity does not suggest that one can arbitrarily theorize what there is, but rather indicates that different conceptual systems can differ in their commitment to the same reality. Comparing different conceptualization approaches to different measurement systems, Davidson writes, "[we] know there is no contradiction between the temperature of the air being 32° fahrenheit and 0° celsius; there is nothing in this 'relativism' to show that the properties being measured are not 'real'."[47] No matter which system is used to quantify it, temperature is committed to being a quantifiable fact. Both measurement systems are not developed arbitrarily but are commonly recognized methods. This is also why his anomalous monism stresses that the mental and physical vocabularies cannot replace each other. Both physical and mental explanations are objective explanatory systems that undertake commitments to reality. The mental vocabulary used in the everyday attribution of mental states is not subjectively arbitrary, but it reflects the commitment of the linguistic community to the shared reality. Accordingly, the mental events presented by the mental explanatory system are as objective as the physical events presented by the physical explanatory system.[48]

One could counter that this claim does not address the explanatory gap in the mind–body problem, but rather turns the different explanatory frameworks into compatible and complementary relationships. But the recognition of different explanatory systems undertaking commitments to reality demonstrates precisely that different explanatory systems can be placed in a broader logical space that allows interpretations to complement each other. Interpretations can reasonably complement each other not because the objects of interpretation are committed to physical or psychological objects, but because the interpretations are carried out consistently and coherently. This returns to Davidson's advocacy of the principle of charity as interpretative strategy, where using one conceptual system to interpret expressions in another conceptual system requires first presuming them to be interpretable in a correspondent and coherent manner. Hence Davidson suggests that the point is not to find some strict psychophysical laws that unify physical and mental explanatory frameworks, but to allow different explanatory frameworks to hold true to their respective evidence.[49] He concludes that "nomological slack between the mental and the physical is essential as long as we conceive of man as a rational animal."[50]

---

[47] Donald Davidson, "What Is Present to the Mind? [1989]" In *Subjective, Intersubjective, Objective* (New York: Oxford University Press, 2001), 65.

[48] Davidson emphasizes that "anomalous monism does not suggest that mental events and states are merely projected by the attributer onto an agent; on the contrary, it holds that mental events are as real as physical events, being identical with them, and attributions of states are as objective." Davidson, "Indeterminism and Antirealism," 72.

[49] Davidson, "Mental Events," 222.

[50] Ibid., 223. Significantly, Davidson notes that the dichotomy between psychological and physical systems of explanation is that of teleology and causation. He cites Kant's treatment of the antin-

To conclude, the translation indeterminacy thesis illustrates, on the one hand, that understanding of meaning is an interpretive relation between subjects with respect to the truth conditions. This starts from the interpreter's rational comprehension by implementing the principle of charity, allowing the speaker's expression to be taken as meaningful and the speaker to have true beliefs. On the other hand, translation indeterminacy reveals the relativity and commitment of ontology in a semantic ascendant sense. Meaning, both in terms of meta-linguistic level interpretability and in terms of presenting facts, is grasped on the basis of the commitment to a reality that can be presented by linguistic structures. Grasping meaning thus presupposes recognition of the underlying patterns that constitute such a reality, which can be interpreted in different ways. The next subchapter will reveal how interpretable patterns are formed and perceived in shared practices, to provide available information for interacting agents.

4.3 Patterns of Meaning-Making in Triadic Interaction

In the interpretivist approach outlined above, meaning, truth, and belief are interrelated. Interpreting an expression requires first adopting such an attitude toward it that it could be true under certain conditions explainable by the metalanguage. Therefore, interpreting meaning can be understood as a linguistic practice that draws on recursive structures, which commit interpretable patterns for linguistic practitioners. This subchapter is dedicated to further exploring the function and formation of interpretable patterns. After outlining Davidson's externalized epistemology, it will argue that the triadic interaction in the modified Turing test is both an access for the interrogator to acquire knowledge about others and a condition for the test subject to develop action patterns that can be attributed to thinking.

4.3.1 Inter-Objective Foundation of Knowledge

Ontological commitment and relativity reveal not only the commitment to what there is, but also involve the epistemological question of how to recognize it. If there is indeed something, then it should be possible for more than one person to recognize it, albeit from different perspectives. This assumption of universal validity of recognition presupposes the common ability of humans as rational agents. How is this ability exercised? Here it is helpful to distinguish between two paths: Quine's naturalistic epistemology and Davidson's externalized epistemology. The

omy regarding free will and universal causality, noting that the recognition of incompatible explanatory frameworks is a necessary condition for human beings to act as rational agents. (225)

naturalistic approach focuses on internal cognitive processes, while the externalized approach emphasizes the roots of knowledge formation in action and interaction.

The basic epistemological problem is, in Quine's view, the problem of how to acquire knowledge.[51] He believes that all human knowledge originates from physical stimuli, with daily empirical knowledge being the product of sensory systems responding to stimuli, while scientific knowledge derives from empirical data induced by stimuli.[52] Following this line, he attempts to uncover how we acquire confirmable, coherent knowledge about the world starting from stimuli. He devoted himself to developing holistically coherent scientific theories from the standpoint of confirmatory holism.[53] What he calls science is a very broad concept that includes not only the natural sciences, but also disciplines such as psychology, economics, sociology, etc.[54] He argues that these disciplines can generate empirical knowledge through similar scientific methods and procedures, such as proposing hypotheses and then evidencing them.[55]

The approach to epistemological issues centered on scientific methods is also called naturalized epistemology, which does not assume a thoroughly irreconcilable conflict between scientific and everyday knowledge. Empirical knowledge acquired through scientific methods is regarded as an extension of stimulus-based common sense. Both scientific and everyday knowledge are adjustable theories in the web of belief. Naturalized epistemology has broadly influenced the study of cognition, explaining cognition in scientific terms and analogizing everyday reasoning with the construction of scientific theories. One of the most prominent theses influenced by this is the assumption that the construction of scientific theories and children's cognitive development are very similar.[56] According to it, both children and scientists understand the world through theoretical hypotheses and empirical evidence based on the stimuli they receive, and the web of belief that serves as the model of understanding can be continually modified in light of experience.

---

[51] Quine, *Pursuit of Truth*, 27.

[52] Ibid., 94.

[53] To achieve holistically coherent theories, he suggests that different disciplines can adopt similar scientific approaches to constructing theories, because "it is within science itself, and not in some prior philosophy, that reality is to be identified and described". Willard V. O. Quine, *Theories and Things* (Cambridge, MA: Harvard University Press, 1981), 21.

[54] Willard V. O. Quine, *From Stimulus to Science* (Cambridge, MA: Harvard University Press, 1995), 49.

[55] Quine, *Pursuit of Truth*, 73.

[56] Alison Gopnik, "The Scientist as Child," *Philosophy of Science* 63, no. 4 (1996): 485. This thesis assumes that both children's cognitive development and scientific theory development involve abstract, coherent theories that are used to predict and explain events. The view that both ordinary cognition and the scientific method derive from the sensory basis of receiving stimuli can be traced back to naturalistic account of epistemology. Gopnik writes: "The idea that science is related to our ordinary cognition, and that both science and ordinary cognition work for evolutionary reasons is not, of course, new. It is the basic idea behind the "naturalistic epistemology" of Quine and others." (489)

However, naturalized epistemology, which attempts to describe the acquisition of empirical knowledge using the scientific method, cannot account for why empirical knowledge is reasonable. Naturalized epistemology takes empirical data as the source of theoretical knowledge, but why and in what sense experience is reliable is precisely the question that traditional epistemology investigates. Indeed, naturalized epistemology focusing on describing knowledge acquisition does not take over the tasks of traditional epistemology. Davidson notes that naturalized epistemology bypasses the key to epistemology, which is to ground the legitimacy of knowledge.[57] Obviously, naturalized epistemology presumes a naturalistic position that turns the task of epistemology into causal explanations of knowledge acquisition, but this naturalistic explanatory framework does not explain its own legitimacy.

To explore the foundations of knowledge and its legitimacy, Davidson focuses on the roots for objective knowledge in shared practices.[58] He proposes that epistemic issues should not be naturalized, but rather externalized into the mutual recognition relations and shared commitments that subjects undertake for objective knowledge.[59] Merely exploring how stimuli become knowledge only gives a third-personal account of the subject's ability to perceive. It cannot reveal how the perceiver forms true beliefs that can be justified. We have seen in the previous subchapter that the belief that an expression can be true is the starting point for interpreting meaning. This belief is also the initial attitude toward the expression of others, allowing for a concept of objective truth that is commonly recognized with others. Davidson states that acquisition of propositional knowledge "requires possession of the concept of objective truth, and this concept is accessible only to those creatures in communication with others."[60] This is the basic idea of externalized epistemology, taking the commonly recognized concept of objective truth in traidic interactions as the foundation of interpretable empirical knowledge. In this sense, knowledge that can be

[57] He writes: "Quine has proposed that epistemology be naturalized. By this he has meant that philosophy should abandon the attempt to provide a foundation for knowledge, or otherwise to justify it, and should instead give an account of how knowledge is acquired." Donald Davidson, "Epistemology Externalized (1990)," In *Subjective, Intersubjective, Objective* (New York: Oxford University Press, 2001), 194.

[58] Quine compares his disagreement with Davidson on the sources of knowledge: "Davidson proposed providing for intersubjective likeness of stimulation by locating the stimulus not at the bodily surface but farther out, in the nearest shared cause of the pertinent behavior of the two subjects. Failing a rabbit or other body to the purpose, perhaps the stimulus would be a shared situation, if ontological sense can be made of situations. But I remain unserved in locating stimulation at the neural input, for my interest is epistemological, however naturalized. I am interested in the flow of evidence from the triggering of the senses to the pronouncements of science. My naturalism does allow me free reference to nerve endings, rabbits, and the other physical objects, so I could place stimulus out where Davidson does without finessing any reification on the subject's part." Quine, *Pursuit of Truth*, 41f.

[59] Davidson, "Epistemology Externalized," 194ff.

[60] Davidson, *Subjective, Intersubjective, Objective* (New York: Oxford University Press, 2001), xvii.

taken as true is mediated by shared objects and the environment, i.e., it is inter-objective.[61]

Triadic interactions involve three kinds of knowledge, namely, knowledge about oneself, knowledge about the external world, and knowledge about other minds.[62] These three forms of knowledge are distinct in that knowledge about one's self appears to be unmediated and has the first person authority; knowledge about the world depends on how one's sensory system receives stimuli; and knowledge about other minds seems to have no direct access.[63] Davidson notes that none of these three sorts of knowledge can be reduced to one or two, all three depend on each other and form a consistent network of beliefs based on "perceptual beliefs".[64] Perceptual beliefs do not directly refer to stimuli. The contents of perceptual beliefs are what causes them, such as the objects shared in the interaction. That is, perceptual beliefs undertake the commitment to causal relations between the objects of beliefs and network of beliefs, allowing them to be interpreted in a way that satisfies correspondence. Also, according to the principle of coherence in interpretation, perceptual beliefs need to be consistent with each other:

> The nature of interpretation guarantees both that a large number of our simplest perceptual beliefs are true, and that the nature of these beliefs is known to others. [...] Any particular belief may indeed be false; but enough in the framework and fabric of our beliefs must be true to give content to the rest. The conceptual connections between our knowledge of our own minds and our knowledge of the world of nature are not definitional but holistic. The same is true of the conceptual connections between our knowledge of behavior and our knowledge of other minds.[65]

In this way, perceptual beliefs as the smallest interpretive units of implementing the principle of charity can form a consistent framework of beliefs. Whether it is knowledge about oneself, about the world, or about others, insofar as one sort of empirical knowledge is claimed to be objective, it should be complemented by the other two. This is because objectivity is understood here as a relationship obtained by triangulating one's own and others' responses to their shared world. In Davidson's words, objectivity is "for each person, the relation between his own reactions to the world and the reactions of others."[66] The requirement of objectivity ensures that individuals maintain a relationship of mutual observation and recognition between self and others and with their shared world, respectively. Once a stable three-way interaction

---

[61] The term "interobjectivity" was proposed by Latour to supplement the concept of intersubjectivity. He argued that in social interactions, agents do not necessarily need to be present simultaneously; they can also mediate and organize their social interactions symbolically through objects, especially artifacts. Therefore, relations between human agents mediated by objects can be regarded as interobjective. Bruno Latour, "On Interobjectivity," *Mind, Culture, and Activity* 3, no. 4 (1996): 228–245.

[62] Davidson, "Three Varieties of Knowledge," 205.

[63] Ibid.

[64] Ibid., 206.

[65] Ibid., 213f.

[66] Ibid., 218.

is developed, it is possible to form mutually recognized information processing agents and semantic information.

Before moving into a deeper discussion of triangulation in interaction, it is necessary to further clarify the significant relationship between perceptual beliefs and interpretable patterns in knowledge acquisition and justification. In both naturalized and externalized epistemology, the source of knowledge is understood as received empirical data. The latter highlights how the concept of objective truth that knowledge entitles can arise when different individuals acquire empirical data in different ways. What is externalized, therefore, is the concept of truth undertaken by knowledge, which is not in the mind of the knower nor in her brain, but in a logical space shared with others. The foundation and justification of knowledge are then shifted from the natural realm to the logical realm shared between subjects. As truth is put into the logical space, reasons for justifying a belief also need to be in the logical space.[67] Believing a proposition to be true in naturalized epistemology is a theory that needs to be proved by science, whereas in externalized epistemology beliefs are understood as doxastic attitudes toward truth.[68] The externalized concept of truth requires beliefs to be interpreted consistently and correspondently, in a way that another rational agent can also understand. Thus, empirical knowledge based on perceptual beliefs cannot be reduced to natural scientific explanations, but it involves normative dimensions of mutual recognition and constraint between subjects. Perceptual beliefs carry commitments to external causes that trigger them and give reasons for intelligible interpretations.

But what exactly are beliefs? In his essay *Real Patterns*, Dennett asks about the nature of beliefs and overviews several representative views.[69] One prevailing view is representational realism, in which beliefs are considered representations in the mind while the brain is the representational system.[70] In Sect. 2.2.2, we have seen that Fodor's representational account holds that mental representations have a language-like syntactic structure and that beliefs can then be formulated as propositional attitudes. Such syntax-driven mental representations are computational processes realized in the brain. However, if beliefs are considered as neural representations, then beliefs are reduced to some functional entities that exist in the brain and body. This mistakenly treats beliefs as physical entities. Another view is to treat beliefs as behavioral dispositions, as Ryle suggests. Complex dispositions

[67] As mentioned in Sect. 3.1.1, in the JTB tradition of epistemology, truth and justification are detached. Although knowledge is treated as a justified true belief, truth corresponds to an objective causal explanation, while the reasons used to justify it are subjective coherent explanations. In externalized epistemology, in contrast, the concept of truth and its justification take place in interpretation. Although Davidson's focus is on interpretations that guarantee causal and reasoned explanations, a broader logical space is evidently needed here to guarantee their cohesion. I will delve into the discussion on logical space of reasons in Chap. 5.

[68] Robert B. Brandom, *Articulating reasons. An Introduction to Inferentialism* (Cambridge, MA: Harvard University Press, 2001), 146.

[69] Daniel C. Dennett, "Real Pattern," *The Journal of Philosophy* 88, no. 1 (1991): 27.

[70] Ibid., 46.

cannot simply be reduced to physical processes that have occurred but should be specified in terms of semi-hypothetical propositions. The indeterminacy of translation demonstrates the very character of complex behavioral dispositions that can be theorized in different ways. Behavioral dispositions are thus committed realities that can be actualized according to different conditions. The commitment to realizability rather than to an external reality leads us to understand behavioral dispositions within the framework of an interpretivist account. Dennett endorses Davidson's interpretivist analogy between beliefs and measurements, indicating that beliefs are not reality per se, but "indirect 'measurements' of reality diffused in the behavioral dispositions of the brain (and body)".[71] That is, beliefs can be understood as practical strategies that can either be expressed explicitly as theories about reality or performed in ways that implicitly know how to respond to reality. Accordingly, perceptual beliefs in interaction are strategies that guide interaction, making sure that there are patterns of action and interaction that can be recognized and interpreted.

The benefit of treating beliefs as different ways of realizing dispositions is that a perceptual belief is not only a sensory certainty, but also provides evidence for other beliefs. Sensory certainties are thus not isolated receptions and representations of stimuli, but mutually evidenced and explained in a holistic web of belief. Thus, what triggers perceptual beliefs can be attributed to some external cause, and the explanations for such causes can be varied. What is important here is that perceptual beliefs commit to patterns that are available for explanation. As Davidson notes, "[t]he value of the patterns is that they reduce a vast, by us indescribably complex, physical situation to something we can grasp, and something on the basis of which we can make rough predictions."[72] Perceptual beliefs and pattern recognition are a grasp of the same cognitive process. Perceptual beliefs commit to its presence while patterns are the way it is realized. The word pattern, in its most basic and literal sense, means repeated or regular occurrence. This meaning is embedded in the concept of pattern recognition, where what can be recognized must be capable of recurrence. Recognizing recurrence requires interpretable recursive structures carried by perceptual beliefs.[73] The complex behavioral dispositions of oneself and the other in interactions can be perceived as patterns that can be used to attribute and predict actions. How these patterns are formed in the interaction and serve conceptual thoughts will be elaborated in the next section.

---

[71] Ibid.

[72] Davidson, "Indeterminism and Antirealism," 81.

[73] Dennett argues that mental states such as attitudes can be treated as patterns, and that these patterns can be interpreted by the observer to perceive and predict behavioral tendencies. In this case, beliefs and other attitudes are not committed to objective reality but are theoretical artifacts when one adopts what Dennett calls "intentional stance" for interpretation. Cf. Daniel C. Dennett, *The Intentional Stance* (Cambridge, MA: MIT Press, 1987), 15ff. This view of patterns and intentions may be seen as an instrumentalist statement. But a narrative based on an interpretive stance is not a fiction but can be understood as an operationally effective theorizing, and accordingly the choice of stance is not arbitrary but is conditioned by practical experience.

4.3.2 The Formation of Pragmatic Action Patterns Through Triangulation

We have seen a critical shift in the externalization of knowing, where justifiable true beliefs are extended from individual perception to intersubjective recognition. The foundation of empirical knowledge lies not only in the reception of sensory data by individuals, but in patterns that can be recognized in different ways committed by perceptual beliefs. Externalized epistemology based on the concept of objective truth treats recognizable patterns as publicly accessible evidence that affords consistent and coherent interpretations of the actions of rational agents. So how are these patterns generated and used in interactions? Davidson proposes triangulation as the basic method for forming and recognizing interpretable patterns. Triangulation arises in the setting of two agents interacting with each other and their simultaneous interaction with a common world, where each agent is related to the other and to the shared world. In his words:

> The basic situation is one that involves two or more creatures simultaneously in interaction with each other and with the world they share [...] It is the result of a threefold interaction, an interaction which is twofold from the point of view of each of the two agents: each is interacting simultaneously with the world and with the other agent.[74]

Each agent in the threefold interaction engages with the world and the other at the same time. This allows the agents to gain experience about the world, about each other, and about themselves. In triadic interaction, agents can generate and interpret recognizable patterns through triangulation to gain empirical knowledge committed to objective truth. Davidson thus sees triadic interaction as a precondition for generating discernable semantics and conceptual thoughts.[75] This is the reason why he takes the triadic interaction as the ground of the test in his modification of the Turing test.

Specifically, the triadic interaction enables an agent to recognize that she and the other are responding to stimuli from a certain direction in their shared world from different standpoints. Considering one's own and the other's responses from different standpoints is triangulating the stimuli from different perspectives.[76] When an agent perceives stimuli from one direction along with the response of the other, she can correlate the stimuli that causes the response of the other with the stimuli that she receives and attribute them to a common cause. The perception of the other agent's response is necessary for the assumption of a common cause, while the perception of stimuli from the same direction is also necessary for the assumption that the other's response is relevant. In this way, the attributed common cause not only provides a causal explanation, but also suggests that it must be a cause that can

[74] Donald Davidson, "The Emergence of Thought [1997]," in *Subjective, Intersubjective, Objective* (New York: Oxford University Press, 2001), 128.

[75] Ibid., 129.

[76] Davidson, "Three Varieties of Knowledge," 213.

trigger similar reactions in the other and in oneself so that their reactions can be perceived by the other. Namely, an attributed common cause is not only a cause to which an agent reacts, but also serves as a reason to ensure that one's own and the other' s responses are relevant. Common causes involve both a physical explanatory framework as well as a normative interpretive framework that considers the other' s perspective. Davidson gives an example of a child and an adult responding to a table:

> It is a form of triangulation: one line goes from the child in the direction of the table, one line goes from us in the direction of the table, and the third line goes between us and the child. Where the lines from child to table and us to table converge, 'the' stimulus is located. Given our view of child and world, we can pick out 'the' cause of the child's responses. It is the common cause of our response and the child's response.[77]

In this case, both the child and the adult respond to the direction of the table, both perceive each other's responses and gain beliefs about the stimuli and each other. One can take one's own standpoint and the perceived standpoint of the responding child as evidence pointing to a common cause in the shared world. Perceptual beliefs about the common cause are confirmed by perceptual beliefs about the other's response, and perceptual beliefs about the other are confirmed by perceptual beliefs about the stimuli attributed to the common cause. As such, perceptual beliefs are both ensured to be caused by stimuli and keep interpretive relationships with other beliefs.

To develop perceptual beliefs in triangulation is to generate interpretable patterns in the experience of interaction with others and the shared environment. Patterns that can be recognized and interpreted in interaction arise not only from stimuli in the environment, but also from other interactants. The latter can be termed as pragmatic action patterns, which are committed by perceptual beliefs about others in triangulation as publicly accessible information.[78] The meaning of these information patterns is contributed by the interaction and its context, i.e., they possess semantic properties in a pragmatic sense. Experienced agents in interaction develop specific ways of acting on the one hand and can recognize the action patterns of others on the other. These patterns enable agents to act consistently on their own, as well as enable them to recognize and attribute how and why others act. Obviously, the process of generating action patterns and developing related beliefs implements the principle of charity, namely, treating the causes and reasons for the other's reaction and action as understandable and reasonable. More specifically, action patterns are treated as if they can be interpreted as carrying coherent beliefs and as being consistent with reality. The principle of charity in triangulation treats not only the

---

[77] Donald Davidson, "The Second Person [1992]," in *Subjective, Intersubjective, Objective* (New York: Oxford University Press, 2001), 119.

[78] Schumann and Du introduced the concept of pragmatic action patterns in a paper on modeling human-machine interaction in triadic interaction. Pragmatic patterns of action there refer specifically to forms of action acquired by experiences of social practices that are framed by materialized social environments. These patterns are produced in social interaction and stabilized as affordances for social cognition. Cf. Nadine Schumann and Yaoli Du, "Machines in the Triangle, a Pragmatic Interactive Approach to Information," *Philosophy & Technology* 35, no.32 (2022): 10ff.

actions of the other as rational, but also the shared environment as affording the possibility of rational action for both interactors. Interactors thus share not only the environment in which they live, but also a coherent and meaningful context consisting of the committed interpretable patterns. And it is the world shared by the agents in the triadic interaction.

The shaping of interpretable patterns through triangulation to construct a shared meaningful world is a key condition for developing conceptual thoughts with objective truth. This is because shaping and recognizing patterns in triangulation allows for constant adjustment of perceptual beliefs about the world, others, and oneself. If the interacting agent allows her own beliefs to be revised, then it is possible for her to distinguish between her own beliefs and concepts carrying objective truth independent of her own beliefs. Davidson considers the ability to distinguish between beliefs and concepts to be an important indicator of thinking:

> A creature that has a concept knows that the concept applies to things independently of what it believes. A creature that cannot entertain the thought that it may be wrong has no concepts, no thoughts. To this extent, the possibility of thought depends on the idea of objective truth, of there being a way things are which is not up to us.[79]

The ability of conceptual thinking requires the agent to acknowledge that there are truths that do not depend on her own beliefs, so that her own perceptual beliefs may be wrong and can be revised. Multiple agents interacting in a shared environment allows the agent to become aware of this and practice triangulating her own beliefs and shared concepts. As indicated above, the addition of a second person in triangulation triggers the principle of charity, providing additional perspective and evidence for the first person's beliefs about a common cause. Namely, triangulation allows an agent to externalize her authority over knowledge acquisition and confirmation into the space she shares with others, the space constituted by a particular perspective and other perspectives. In this space, perceptual beliefs are revisable. As Davidson notes triangulation "creates the space needed for error, not by deciding what is true in any particular case, but by making objectivity dependent on intersubjectivity."[80] The acknowledgement that beliefs are fallible and can be revised suggests that the agent makes a distinction between subjective beliefs and knowledge bearing the concept of objective truth. It also reveals that the acquisition and determination of objective knowledge is intersubjectively constrained.[81] The ability to think conceptually is thus not only to have beliefs about something, but also to recognize that such beliefs may be wrong, and hence to acknowledge that such beliefs may be true only under certain conditions.

Insofar as the three-way interaction continues, it is possible for both agents exercising triangulation to develop mutually recognizable and interpretable patterns of action. Long-term stable triadic interactions imply that the other agent is engaging

[79] Donald Davidson, "What Thought Requires [2001]," in *Problems of Rationality* (Oxford: Oxford University Press, 2004), 141.

[80] Ibid., 143.

[81] Ibid.

in the interaction in a similar way to the extent that the three-way interaction is stable. This suggests that the other agent is also capable of attributing common causes and recognizing action patterns and pursuing knowledge confirmation subject to intersubjective constraints. That is, the other interactor in the long-term interaction should also be capable of triangulation and conceptual thinking. Davidson writes:

> It is clear that for triangulation to work, the creatures involved must be very much alike. They must class together the same distal stimuli, among them each other's reactions to those stimuli. In the end, it is just this double sharing of propensities that gives meaning to the idea of classing things together. […] What makes the responses similar? Our concepts do; we have the concepts that define these classes. It takes another creature enough like the first to see and say this. The sharing of many discriminatory abilities explains why a considerable degree of holism is no obstacle to communication. This is also why Turing had the right idea about how to tell if a device (or animal) is thinking.[82]

Returning to Davidson's modification of the Turing test, it now becomes clearer why he suggests that the test should be set up as a long-term interaction between the human interrogator, the test subject, and their shared environment. The three-way interaction on the one hand enables the human interrogator to adequately interpret whether the test subject's expressions make sense, that is, whether they can be true under certain conditions and whether the test subject has relevant beliefs. On the other hand, the test subject has the opportunity to practice triangulation in the three-way interaction, exercising the ability to acquire objective knowledge that takes into account the standpoint of the other. In a stable triangulation interaction, both interactors can attribute common causes that cause their respective responses, form and mutually recognize patterns of actions and reactions, and develop reasonable interpretations. A stable triangular interaction also signals the interrogator's recognition and affirmation of the test subject's intelligence of a similar rational way of interaction to her own. At the end of Davidson's revision of the Turing test, he writes: "Can the interrogator now tell what the object thinks? The answer is that it depends on how long the interrogator can question and observe the object."[83] By externalizing epistemological questions that include knowledge about others into triangulation, the test of whether someone is thinking becomes an empirical question in a long-term interaction.

As a conclusion, artificial intelligent agents as test subjects can only pass the Turing test if they are allowed to practice and interact with human agents in a shared world. The important implication of the modified test is that unless the foundation of knowledge originates in triadic interaction and intersubjective recognition, there can be no objective knowledge and thus no convincing judgment. By placing the Turing test in a long-term triadic interaction, it is possible for the human interrogator to interpret the conditions under which the test subject's expressions are meaningful and the test subject's beliefs are true. In the modified test, the interrogator no longer observes expressions that can be directly taken as evidence, but together with

---

[82] Ibid.

[83] Davidson, "Turing's Test", 84.

the test subject generates interpretable patterns that compose meaningful contexts to inform judgments. Triadic interaction is also the basic model for generating patterns of semantic information in the sense of data that can be queried about their origins. The delivery and processing of semantic information thus cannot be reduced to a dyadic data exchange between sender and receiver, but concern also the mutual shaping of interpretable information patterns and the corresponding information processing agents. The modified Turing test is accordingly no longer an imitation game for the test subject, but an interplay between the interrogator, the test subject, and their shared world. In the modified version, the question of whether machines can think is no longer answered in terms of their intrinsic capabilities, but in terms of the acquired conditions under which machines can be interpreted as thinking.

So far, the criteria for determining whether the other thinks and the conditions for developing conceptual thinking are externalized to long-term triadic interactions. Shared interpretable patterns afford perceptual beliefs and empirical knowledge about the world, the other, and oneself. Long-term triangulation allows agents to practice specific ways of interacting and acting that can be recognized and interpreted. Davidson notes that interpretable thought requires a proper history of experience that can be observed, as he writes, "if I am right, [an artifact] thinks only if its thinking can be understood by a human interpreter, and this is possible only if the artefact physically resembles a person in important ways and has an appropriate history."[84] He emphasizes that physical resemblance to a person does not mean similarity in origin, building materials and appearance, but rather like a person who can be observed engaging in the world and learning behavioral patterns that can be interpreted.[85] The triadic interaction of agents in a shared world opens up the possibility for the development of an observable history of individual experience. Thus, thoughtful and interpretable AI agents can only be actualized and recognized by being allowed to shape such observable histories of experience in shared practices with humans. A freshly shipped artifact cannot be expected to be as intelligent as a college student with eighteen years of life experience in engaging with others, as Turing pointed out.

Importantly, the individual experience of human agents does not arise in idealized triadic interactions, but in the space of already constructed and institutionalized social norms. Established norms frame human interactions and interpretations. As social practices are expanded by computer and internet technologies, the conventions and institutions of practice are also being implicitly coded into cyberspace. Interconnected cyberspace becomes the shared practice environment for information processing agents, providing both human and artificial agents with accessible

[84] Ibid., 85.

[85] In the other essay Davidson asks the questions, "what we must add to computer to ensure they can think." He first lists candidates such as origins, building materials, and appearance, and argues that they are detachable properties of thought, insofar as they do not inhibit the possibility of communication and interpretation. Then, he points out that history is the key prerequisite for thinking, as artifacts need experience from causal interactions with the world to learn about it. Cf. Davidson, "Representation and Interpretation," 87ff.

information patterns. If these patterns do afford possibility for meaning making and conceptual thought, as argued in this chapter, then information processing agents who can appropriately engage with them are the agents who generate meaning and conceptual thinking. Further, by authorizing information processing agents to share the resources and environments of human practices, artificial agents can engage in triadic interactions with human agents.

Bibliography

Brandom RB (2001) Articulating reasons. An introduction to inferentialism. Harvard University Press, Cambridge, MA

Davidson D (2001a) Epistemology externalized [1990]. In: Subjective, intersubjective, objective. Oxford University Press, New York

Davidson D (2001b) In defence of Convention T [1973]. In: Inquiries into truth and interpretation. Oxford University Press, Oxford

Davidson D (2001c) Indeterminism and antirealism [1997]. In: Subjective, intersubjective, objective. Oxford University Press, New York

Davidson D (2001d) Mental events [1970]. In: Essays on actions and events: philosophical essays, vol 1. Oxford University Press, Oxford

Davidson D (2001e) On the very idea of a conceptual scheme [1974]. In: Inquiries into truth and interpretation. Oxford University Press, Oxford

Davidson D (2001f) Radical interpretation [1973]. In: Inquiries into truth and interpretation. Oxford University Press, Oxford

Davidson D (2001g) Semantics for natural languages [1970]. In: Inquiries into truth and interpretation. Oxford University Press, Oxford

Davidson D (2001h) The emergence of thought [1997]. In: Subjective, intersubjective, objective. Oxford University Press, New York

Davidson D (2001i) The second person [1992]. In: Subjective, intersubjective, objective. Oxford University Press, New York

Davidson D (2001j) Theories of meaning and learnable languages [1965]. In: Inquiries into truth and interpretation. Oxford University Press, Oxford

Davidson D (2001k) Three varieties of knowledge [1991]. In: Subjective, intersubjective, objective. Oxford University Press, New York

Davidson D (2001l) Truth and meaning [1967]. In: Inquiries into truth and interpretation. Oxford University Press, Oxford

Davidson D (2001m) What is present to the mind? [1989]. In: Subjective, intersubjective, objective. Oxford University Press, New York

Davidson D (2004a) Representation and interpretation [1990]. In: Problems of rationality. Oxford University Press, Oxford

Davidson D (2004b) Turing's test [1990]. In: Problems of rationality. Oxford University Press, Oxford

Davidson D (2004c) What thought requires [2001]. In: Problems of rationality. Oxford University Press, Oxford

Dennett DC (1987) The intentional stance. MIT Press, Cambridge, MA

Dennett DC (1991) Real pattern. J Philos 88(1):27–51

Floridi L (2011) The philosophy of information. Oxford University Press, Oxford

Gopnik A (1996) The scientist as child. Philos Sci 63(4):485–514

Gruber TR (1993) A translation approach to portable ontology specifications. Knowl Acquis 5(2):199–220

Latour B (1996) On interobjectivity. Mind Cult Act 3(4):228–245
Quine WVO (1960) Word and object. MIT Press/Wiley, New York
Quine WVO (1969) Ontological relativity and other essays. Columbia University Press, New York
Quine WVO (1981) Theories and things. Harvard University Press, Cambridge, MA
Quine WVO (1992) Pursuit of truth. Harvard University Press, Cambridge, MA
Quine WVO (1995) From stimulus to science. Harvard University Press, Cambridge, MA
Quine WVO, Ullian JS (1978) The web of belief. McGraw-Hill, New York
Rorty R (1986) Pragmatism, Davidson and truth. In: Lepore E (ed) Truth and interpretation. perspectives on the philosophy of Donald Davidson. Blackwell, Cambridge, pp 333–355
Schumann N, Du Y (2022) Machines in the triangle, a pragmatic interactive approach to information. Philos Technol 35(32). https://doi.org/10.1007/s13347-022-00516-4
Tarski A (1936) The concept of truth in formalized languages. In: Tarski A (ed) Logic, semantics, metamathematics. Oxford University Press, Oxford, pp 152–278

Chapter 5
Rational Agency in the Normative Space of Reasons

This chapter will introduce the normative inferentialist approach developed by Sellars and Brandom, which accounts for how the logical and normative space of reasons is constituted in discursive practices together with the mutual authorization of rational agents. This conception of rational agents can be fruitfully applied to attribute and license the normative status of non-human rational agents. As will be argued in this chapter, the space of reasons shared with humans allows rational agents to be characterized by their normative statuses instead of mental states. According to this approach, interpretable and responsible artificial rational agents do not mean turning the black box of inner operations into a glass box but rather placing the formation and performance of information processing procedures into the space of reasons. Because the conceptual framework of normative inferentialism elucidates how rational agents can make implicit commitments explicit in norm-making through the expressive role of linguistic structures. This conceptual framework will be employed to specifically reveal how information patterns formed through digital practices and processed by semantic technologies enable both human and artificial agents participating in digital social practices to undertake discursive commitments and achieve rational action.

5.1 In-Formation of the Normative Space of Reasons

When Sellars suggests that the logical space of reasons is where knowing takes place, he indicates that putting thought episodes or states into the space of reasons also authorizes that they can be justified.[1] Expressing the corresponding sensory experience in terms of "it is green" does not merely provide a report or a description

[1] He wirtes: "The essential point is that in characterizing an episode or a state as that of knowing, we are not giving an empirical description of that episode or state; we are placing it in the logical

Y. Du, *Machines in Minds*, Studies in Applied Philosophy, Epistemology and Rational Ethics 78, https://doi.org/10.1007/978-3-032-15842-0_5

of color perception, but also holds a commitment that what is being expressed is justifiable. Insofar as a thought episode or state such as a perceptual belief can be characterized as knowing, then it should be able to be justified. Otherwise, it cannot be recognized as knowing. Brandom calls this view the inferential functionalist approach to semantics, in which semantic content can be known on condition that it can be articulated as a part of inference.[2] But what exactly is inference? For inferential functionalists, inference is not merely a matter of following rules, but always implies a commitment to the normative origins of those rules and a recognition of rational agents who are entitled to make inferences.

5.1.1 Encoding Material Inference in Language Use

Inference, in its original sense, refers to the logical step of moving forward from premises to consequences. In classical logic, the unit of inference is the sentence rather than the constituent parts of the sentence such as words and combinations of words. In syllogism, for example, premises and conclusions are sentences. Accordingly, a common approach to the analysis of everyday inference is to extend everyday linguistic expressions into sentences in syllogism. But does inference in everyday language use really need to follow the steps of strict formal logic? Apparently not. One does not normally need to extend inferences in everyday language use with additional premises and deductions, such as extending "it's raining, therefore the street will be wet" into the subjunctive conditional form "If it were the case both that every time it rains, the streets are wet and that it's raining, then the streets would be wet".[3] In his essay *Inference and Meaning*, Sellars argues that in everyday language use it is not always necessary to bear in mind the rules of formal inference, but rather to follow material rules of committed inference.[4]

---

space of reasons, of justifying and being able to justify what one says." Sellars, "Empiricism and the Philosophy of Mind," 298f.

[2] Brandom, *Brandom reads Sellars*, 124. Brandom remarks: "The contribution to this argument of Sellars's inferential functionalism about semantics lies in underwriting the claim that for *any* judgment, claim, or belief to be contentful in the way required for it to be *cognitively, conceptually*, or *epistemically* significant, for it to be a potential bit of *knowledge* or *evidence*, to be a *sapient* state or status, it must be able to play a distinctive *role in reasoning*: it must be able to serve as a *reason for* further judgments, claims, or beliefs, hence as a *premise* from which they can be *inferred*" (124).

[3] The subjunctive conditional is not the only form of extension, but it is the least presupposed in purely logical form. For a discussion of the problem of extension into other logical forms see Wilfrid Sellars, "Inference and Meaning," Mind 62, no. 247 (1953): 324ff.

[4] To illustrate what the material rules of inference are, Sellars draws on Carnap's distinction between logical and extra-logical transformation rules. Carnap characterizes rules of transformation between sentences generally as rules of inference, which can be classified into logically valid rules of inference and physically valid rules. The former relies on the formal syntax of sentences, while the latter relies on the descriptive content of terms in sentences. In this distinction, formal inference and inference between descriptive contents are independent of each other. However, Sellars points out that this distinction ignores the very nature of inference. Ibid., 313ff.

Sellars argues that rules of inference, whether formal or content-based, involve norms and associated terms such as *should*, *may not*, etc., that force the inference to be performed.[5] If the imperative nature of inferential rules "for *doing* something" is ignored, then inference can be mistaken for automatically following established formal rules or arbitrarily following personal desires.[6] In linguistic practices, the rules for doing something are codified in the expression of doing something with words. This view is in line with Ryle's analysis of knowing how, suggesting that knowing what a descriptive expression means requires first knowing how to use it as a dispositional proposition. Accordingly, rules of inference are in fact norms of how to do things with linguistic symbols, and their formulation from a meta-linguistic level must contain law-like instructions with normative force. Based on this general consideration of inference rules, Sellars examines the commonly used expressions involved in inference in everyday language and finds that even the use of descriptive expressions implies modal logical terms such as *necessary* or normative terms such as *should*. For example, the use of the word *red* implies that *red* as a symbol can correctly convey the meaning of red. Therefore, what appears to be content-based inference must also comply with certain formal rules of symbol use and transformation, rather than semantic rules that are deliberately distinguished from syntactic rules. What is followed in the transformation between descriptive expressions is called by Sellars the material rules of inference, which "determine the descriptive meaning of the expressions of a language within the framework established by its logical transformation rules".[7] Inferences based on the material transformation rules can be expressed explicitly in logical, modal and normative terms, although they are usually implicit in the use of descriptive terms. The validity of the material rules of inference is not authorized through formal rules, but they can only be expressed through formal inferential terms.

Indeed, the material rules of reasoning can be characterized as the enforcement of inferential moves from conditions to consequences between conceptual contents conveyed by descriptive expressions. In Brandom's reading, the symbol of a descriptive expression such as "red" is given meaning in the sense of conceptual content by playing a role in material inference, and the use of this expression carries the commitment to this rule of inference.[8] In other words, material inferences are codified in language practice in the use of descriptive expressions that convey conceptual content. To clarify the rules followed by the transformation between descriptive expressions, one can use logical terms to express implied premises and consequences,

[5] Ibid., 329.

[6] Ibid.

[7] Ibid., 331.

[8] The notion of material inference is interpreted by Brandom as "inferences whose propriety essentially involves the nonlogical conceptual content of the premises and conclusions". Material inferences are thus non-monotonic. Robert Brandom, *Making it Explicit: Reasoning, Representing, and Discursive Commitment* (Cambridge, MA: Harvard University Press, 2001), 101. Harvard University Press, 1994, 1st Ed. Hencefort *MIE*.

especially in the form of subjunctive conditionals.[9] That is to say, the endorsement of material inferences is implicit in the use of descriptive expressions in everyday language, enabling language users to make content-based inferences. By being expressed in subjunctive conditionals, the possible steps of material inference and the default endorsement can be made explicit.

Sellars suggests that the justification of material inferences can be placed in a linguistic game of giving and asking for reasons between interlocutors. To be specific, in the linguistic practices, the interlocutors give and ask each other the reasons for their respective expressions. In this process, the conceptual content conveyed by an expression must undertake an inferential role, which means that it can be used as a premise or conclusion for reasoning. To force expressions to undertake the inferential role is to bring discursive practices under rational control. For by giving and asking for reasons, the implied rules of inference and the endorsement of them will be elaborated and will no longer be somehow privately presupposed. And by allowing interlocutors to constrain each other, rational control prevents the rules of inference from being endorsed only individually. In this way, if an inference can be jointly approved in the game of giving and asking for reasons, then it is a proper inferential move.

Remarkably, the concern with material inferences uncovers the normative origins of the enactment of and the compliance with formal rules. This concern can be traced back at least to Frege's criticism of Boole's symbolic logic for dealing only with the formal part of language, as discussed in Sect. 1.1. Brandom indicates two paradigmatic features of Frege's approach: The first is understanding "the content of nonlogical concepts in terms of their inferential role, rather than in terms of their extensions", and the second is understanding "the task of logical vocabulary to be expressing explicitly what is implicit in those material conceptual contents.[10] These two features specify conceptual content in terms of its inferential role and logical vocabulary in terms of its semantic expressive role, respectively. Both features presuppose that material inferences are necessarily committed to be "prior to the introduction of specifically logical vocabulary and form".[11] Accordingly, material inferences are supposed as conditions for formal inferences that can be expressed by logical terms and forms, especially conditionals. Or to put it differently, what is explicitly expressed by the logical terms and forms is the process of material inferential moves. Therefore, the task of logic can be regarded as making explicit the

[9] He writes: "The material inference codified in subjunctive conditionals are inferential involvements that are essential to the content of the concepts used in science and everyday life. These are not logically valid inferences. But logical vocabulary, subjunctive conditionals, can be used to express these material inferential relations. Without such vocabulary, the inferences can still be endorsed. With it, those content-generating inferential endorsements can be made explicit as the content of a claim or propositional endorsement." Ibid., 103f.

[10] Ibid., 109. He comments that "Frege's overall project for his *Begriffsschrift* is to use conditionals to make it possible to *say* explicitly what the inferential role of ordinary, nonlogical concept is." And the role of conditionals is "making follows from what explicit, as itself a judgeable content, one that can itself appear as a premise or conclusion in inference" (Ibid.).

[11] Ibid., 110.

material inferences implicit in practice into principles that can be expressed by formal inferential steps.

Understanding logic in terms of its expressive role for material procedures begins with the acknowledgement that material procedures can be expressed through logical terms. This acknowledgement is by default implicit in language use. Using a descriptive expression *x* that conveys conceptual content, for example, involves an acknowledgement that under certain conditions the description *x* is appropriate. And logical terms and forms, especially conditionals, can make these conditions explicit. The advantage of treating conditionals as the basic form for explicit expression is that that they can clarify that the necessity of taking something as true is the starting point for formal inference. As explained in Sect. 4.2, taking *x* to be true is the starting attitude for interpreting *x*. Conditionals can express both the attitude of taking *x* to be true and a valid inferential form that brings "if *x* (is true)" as a necessary premise into further logical inference.[12] Namely, the affirmation of *x* as a judgeable pattern that can be true permits the next step of inference. This affirmation needs to be preserved as an implicit commitment in any formal inference that follows. The maintenance and refutation of commitments in the game of giving and asking for reasons is also a process of codifying possible material rules of inference into linguistic practice, that is, the process of in-formation that brings material disposition into logical forms. The expressive role of logic is thus necessary in linguistic practice for the ordering and interpreting of material rules of inference. Accordingly, Brandom proposes an inferentialist semantics that views the meaning of expressions as the "material inferential commitment one who uses them undertakes: the commitment to the propriety or correctness of the inference from their circumstances to their consequences of application."[13]

Why is it important to focus on the normative origins of rational practices? Because normative origins in practice can provide a coherent account of the material and formal rules of inference. Sellars distinguishes between two typical images of humans and the world they inhabit: One is the scientific image, which explain humans and the world by the scientific explanatory framework of how they work in their mechanism of activity; the other is the manifest image, which is a more primitive apprehension of "man-in-the-world" that begins with perception.[14] These two images provide different conceptual frameworks for the connection between actions. Actions in the manifest image can be predicted based on a person's habits and normative constrains which are modified by oughtness, while the activities in the

[12] Ibid., 112.

[13] Ibid., 131. He distinguishes the weak inferentialist thesis from the strong one. "The weak inferentialist thesis is that inferential articulation is *necessary* for specifically conceptual contentfulness." By contrast, the strong inferentialist thesis emphasizes "that broadly inferential articulation is sufficient for specifically conceptual contentfulness-that is, that there is nothing more to conceptual content than its broadly inferential articulation". Brandom himself holds the strong inferentialist position.

[14] Wilfrid Sellars, "Philosophy and the Scientific Image of Man," in *Science, Perception and Reality* (London/New York: Routledge & Kegan Paul/The Humanities Press, 1963), 10.

scientific image follow causal laws and are modified by necessity. Sellars notes that unlike a de-personalized conceptual explanatory framework in the scientific image, the most basic category in the manifest image is "person", who usually has a certain character and whose behavior is predictable to some extent.[15] Such predictions are distinct from causal explanations in the scientific image. The distinction between predictability and causality is related to the difference between "ought" and "is". Rational agents, regardless of the practice community they come from, can use causal explanations or descriptions to make sense of the world. How is this law-like universal nature ground in the space of reasons? This question will be addressed in the next section, together with the unique linguistic and epistemic powers of rational agents.

5.1.2 *Decoding Rationalist Commitments to Modality and Normativity*

In Hume's reflections on classical empiricist philosophy, he proposed that causality could be understood as a kind of mental "constant conjunction" based on customs generated by social practices.[16] For causality itself cannot be directly perceived or noticed in the same way as objects or qualities, but rather as a tendency to grasp the connections between perceived things. Yet, if causality is merely custom-based inertial thinking, then different communities could form different rules, and causality would not have the kind of necessarily law-like nature. Therefore, the universal necessity that distinguishes causality from other customs must be more than mere mental conjunction. Regarding this issue, Brandom compares what he calls the Hume-Quine thesis and the Kant-Sellars thesis, and clarifies a core epistemic condition for language use by rational agents.

Brandom notes that both Hume and Quine, in their reflections on empiricism, focus on the necessary role of rules or theories to explain facts.[17] What they share is the assumption that one can first acquire descriptive representations through perception and then organize them through law-like mental conjunctions or theories at the meta-linguistic level. This means that one' s observation and theorizing are two separate actions. Descriptive representations can be justified by observable evidence, but the logical relations between them, especially induction from the particular to the general, cannot be proved. Accordingly, the validity of causal relations cannot be demonstrated. Brandom notices that a consequence of this line of thinking

[15] Sellars stresses that predictable is not the same as being caused, since "predictable no holds barred, but predictable with respect to evidence pertaining to what the person in question has done in the past, and the circumstances as he saw them in which he did it." Ibid., 11.

[16] David Hume, *An Enquiry concerning Human Understanding* (Oxford: Oxford University Press, 2007), 30ff.

[17] Robert Brandom, *From Empiricism to Expressivism: Brandom reads Sellars* (Cambridge, MA: Harvard University Press, 2014), 145.

is that the validity of modal terms such as *should, could, might*, etc. is also denied because the hypothetical judgments they express cannot be empirically justified.[18] As a result, those who hold to the Hume-Quine thesis either need to explain modality in non-modal terms, or abandon the use of modal terms.

However, modal terms have an irreplaceable role in language use because modality incorporates the attitudes of language users and the norms of language use. A thorough separation of descriptive and modal expressions misunderstands both the role of modal terms and the origin of descriptive expressions. In contrast to the Hume-Quine thesis, Brandom proposes a Kant-Sellars thesis, which states that the use of descriptive expressions does not directly reflect sensory experience but is based on the mastery of modality-based conceptual schemas for perception. The main point of the Kant-Sellars thesis can be briefly stated as follows:

> It is the claim that in being able to use nonmodal, empirical descriptive vocabulary, one already knows how to do everything one needs to know how to do in order to deploy modal vocabulary, which accordingly can be understood as making explicit structural features that are always already implicit in what one *does* in describing.[19]

According to this thesis, even the most basic descriptive expressions can only be used appropriately if they are based on the mastery of conceptual relations that can be made explicit with the help of modal terms. As Sellars notes, the acquisition of knowledge that appears to be obtained through direct observation actually also needs the mastery of a general form of knowledge.[20] In uttering the expression "it is green" one not only expresses the mental representation of green as a sensory experience, but also endorses that it is appropriate to use the expression at this point to refer to that experience, that is, one endorses the normative use of the expression. The ability to refer to what is perceived with appropriate expressions, although it can be understood as an inductive ability built on habits of language use, is not a formal induction but rather allows for experience to be expressible. This is in line with what was discussed in the previous section, namely that the use of descriptive terms inherently involves the commitment to implicit material inferences, rather than following additional logical rules.[21]

Brandom indicates that in using descriptive terms to express empirical knowledge, the language user as a knower necessarily undertakes a "*rationalist* commitment to the necessary inferential articulation of the concepts expressed by the use of such vocabulary".[22] This is the mode of understanding sensory experience that is

[18] Ibid., 149.

[19] Ibid., 143.

[20] Sellars, "Empiricism and the Philosophy of Mind," 299.

[21] In this line of thought, on the one hand, self-evident sensory states or episodes are considered as a "Myth of the Given". (Ibid.) That is, the immediate sensory experience that underlies cognition in traditional empiricism is called into question. On the other hand, if what is known is not directly given, then it is mediated through something. If the mediator is taken to be some given rule, then it falls into the myth of logical given. Cf. Paul Redding, *Analytic Philosophy and the Return of Hegelian Thought* (Cambridge: Cambridge University Press, 2007), 56–84.

[22] Brandom, *Brandom reads Sellars*, 144.

unique to language users, or rather concept users.[23] He notes that the rationalist commitment implicit in the use of concepts is accounted for by Kant, who "saw that all empirical concepts are like their refined descendants in the mathematized natural sciences in this respect: their application implicitly involves counterfactual-supporting dispositional commitments to what *would* happen *if*...".[24] On this view, not only natural scientific propositions that draw on mathematical methods, but also concepts in everyday experience implicitly undertake a dispositional commitment to counterfactual possibilities in the form of "what (would happen) if…". The rational commitment is to allow for the possibility of counterfactuals, i.e., to allow for the legitimacy of the concept to be challenged by making it a link in the inferential articulation. In this way, descriptive expressions take on a dispositional commitment in their use, expressing the initial attitude that it can be true on the one hand, and allowing it to be questioned on the other. Descriptive expression can then be taken as a link in inference both in the form of "as if it were true" or in the form of "what if in some cases it were false". This dual dispositional allows for a universally inferable link between descriptive expressions. Such links are not personal mental associations but are necessary commitments to justifiable normative language use.

Sellars is also concerned with the rational commitment to the necessity of inferential articulation, and he pays particular attention to the role of modal expressions as "*inference licenses,* expressing our commitment to the goodness of counterfactually robust inferences from necessitating to necessitated conditions."[25] Namely, the commitment to the inferential articulation implicit in the descriptive expressions can be made explicit with the help of modal terms, which in turn confirms that the use of descriptive expressions does undertake the commitment to their inferential role. The role of modal expressions here is read by Brandom as entitling descriptive expressions as inferable patterns according to law-like rules.[26] It is the expressive role of modal terms for dispositional commitments that, in a law-like manner, authorizes and enforces descriptive expressions to be used as patterns of inference. Knowing how to use modal terms to confer dispositional commitments on expressions thus lays the groundwork for knowing how to use descriptive expressions appropriately. And knowing how to use descriptive expressions appropriately does not come from mere impressions or observations, but from the necessary rules of use acquired in discursive practices.

---

[23] Ibid., 143. Brandom points out that non-language-using animals do react differentially to various green things without an abstraction notion of green. However, language users understand color in a conceptual way, so that blind and color-blind people can also understand the use of "green" in certain situations, such as traffic lights.

[24] Ibid., 150.

[25] Ibid., 151.

[26] Ibid., 142. He writes: "Sellars's claim, as I am reading him, is that explicitly modal "lawlike" statements are statements that one is committed or entitled to whenever one is committed or entitled to endorse such patterns of counterfactually robust inference, and commitment or entitlement to which in their turn commit or entitle one to the corresponding patterns of inference" (142).

Put together, the Kant-Sellars thesis argues that understanding the dispositional commitments contained in modals is fundamental to the use of descriptive expressions assigned with conceptual content.[27] The use of an intelligible descriptive expression thus entails a modality-based necessary judgement, which imperatively conceptualizes sensory experience as inferable conceptual content.[28] As Brandom points out, "*Judgment* and *agency* are implicitly normative phenomena because they consist in the application of concepts, and applying concepts is undertaking commitments and responsibilities whose *content* is articulated by those concepts".[29] The agency referred to here is the rational agency that commits material inferences through the normative application of conceptual language. This view clearly refutes the Hume-Quine thesis, revealing that it is a mistake to separate descriptive terms from the normative origins that authorize them as patterns of inference. The rational agency attributed to persons is always subject to normative constraints, and the implied norms can be expressed in modal terms. Thus, causality as a law-like universal relation is not a psychological explanation, but is grounded in the recognition of causality as a fundamental category of understanding and judgment.[30] The question of the legitimacy of causal explanation becomes a question of why it is necessary to license law-like inferential relations in conceptualized thinking.

A further implication of the Kant-Sellars thesis is that the rationalist commitment to law-like enforcement places what is intuitively perceived into discursive practices as expressible. That is, the rationalist commitment requires that the reception and processing of data be placed in the space of reasons. To put a perceptual belief into the logical space of reasons is to authorize that it can be understood by

---

[27] Ibid., 150. Brandom remarks: "For Sellars, the laws determining the truth of counterfactuals involving the application of a concept are part of the content of the concept. For Kant, modal concepts make explicit not something implicit in the content of determinate concepts, but something implicit in their empirical use, in applying them to make empirical judgments. That is why the pure concepts of the understanding—what he calls 'categories', such as possibility and necessity—both are to be understood in terms of the forms of judgment (the table of categories derives from the table of judgments) and express synthetic, rather than analytic necessities" (150).

[28] Ibid., 153. Cf. Immanuel Kant, *Kritik der Urteilskraft* (Hamburg: Meiner, 2006), 40, BLIV, IX. According to Kant, the faculty of receiving sense impressions is sensibility, while understanding is the faculty of forming concepts, and rationality [Vernunft] is the faculty of drawing inferences from these concepts. The normative force of judgment is the ability to combine concepts and other thinking episodes into propositions and to hold them to be true or false. It is the ability to assign particular things to general rules, i.e., to decide whether something comes under a certain rule. Thus, the force of judgment provides the link between conceptual understanding and sensory perception. While understanding abstracts concepts as rules from the sense-certainty, the normative force of judgment in turn determines whether these things belong to a certain concept or not.

[29] Brandom, *Brandom reads Sellars*, 167.

[30] Immanuel Kant, *Kritik der reinen Vernunft* (Leipzig: Reclam, 1966), 141. A70 §9. One central idea of Kant's critical theory of epistemology is that knowledge acquisition is mediated by the conceptualization. To conceptualize empirical phenomena is to use categories to acquire empirical knowledge and thereby make judgements. The nature of cause and effect can be understood as four forms of judgement, in his terms, quantity, quality, relation, and modality, and their use depends on concrete empirical practice.

rational agents. What appear to be non-inferential episodes of thought should be regarded as implying material inferential moves that can be expressed and thereby be given reasons, so long as they are thought to be understandable. This line of thought makes the development of explainable rational artificial agents more operational. Because even if artificial agents have different material bases or different sensory experiences than human agents, they can still participate in discursive practices by undertaking rationalist commitments and conceptualizing their experiences. Whether humans or artificial agents, if they can skillfully master how to appropriately license information patterns as inferable patterns and place them in the space of reasons, they are exercising rational agency in discursive practices.

Brandom accounts for the role of authorizing inferential patterns and their normative binding in discursive practices from a broader anthropological point of view. He argues that what distinguishes humans from other animals is the use of concepts in language games.[31] Accordingly, the mutually constraining exercise of using concepts committed to inference in discursive practices lays the foundation for the reciprocally authorized social status of language users. In the process, language users who undertake the rationalist commitment establish the space of reasons and make themselves reasonable beings or rational agents. The linguistic practices of rational beings are thus conditioned and guided by the normative force. He writes:

> Reason is as nothing to the beasts of the field. We are the ones on whom reasons are binding, who are subject to the peculiar force of the better reason.
>
> This force is a species of *normative* force, a rational 'ought'. Being rational is being bound or constrained by these norms, being subject to the authority of reasons. Saying 'we' in this sense is placing ourselves and each other in the space of reasons, by giving and asking for reasons for our attitudes and performances. Adopting this sort of practical stance is taking ourselves as subject of cognition and action.[32]

The central idea here is that by allowing each other to stand in the space of reasons and engage in the game of giving and asking for reasons, discursive practitioners authorize each other as rational agents. Entitled rational agents share the space of reasons and are subject to the normative force of "ought". Thus, rational agents are not merely knowers with cognitive states and processes, but also normatively committed agents who are obligated to justify their knowledge in the space of reasons. The space of reasons is thus neither an individual psychological nor a physical space, but a normative realm that is jointly committed to by rational agents and can be logically articulated. Just as children are brought into the normative space through upbringing and education, nonhuman agents should be allowed into the space of reason if they are expected to become rational. Before we discuss how participants in discursive practices empower each other to enter the space of reasons and exercise their role as responsible rational agents, the next subchapter will first elaborate on the understanding of intentionality within the normative inferentialist framework.

---

[31] Brandom, *MIE*, 4.

[32] Ibid., 5.

5.2 Operative Intentionality in Rational Agency

Intentionality, ever since Brentano proposed it as a key characteristic of mental phenomena, namely as being directed towards an object, has taken a prominent place in discussions about the mind.[33] The attribution of intentional states is also the way rational agents are understood in the manifest image of man, namely, through the conceptual framework of persons with predictable action patterns. Gallagher and Miyahara survey the use of the term intentionality by different philosophers and divide them into two broad groups, "intentionality as character of mental states" and "intentionality as a characteristic of agents".[34] The former views intentionality as an internal cognitive state, the content of which is usually taken to be inseparable from neural activity. In contrast, the latter treats intentionality as an agent's goal-directed way of acting with intention. Both accounts can hardly ground each other in a naturalistic framework. This is because the former focuses on non-derived content and its biological basis, whereas the latter focuses on environmentally conditioned intention in action. Which one is the condition for the possibility of information processing with semantic content? The two accounts of intentionality correspond to cognitivist and behaviorist approaches.

Explaining intentionality within a normative conceptual framework can make the two accounts compatible. In the inferentialist semantics, semantic content is put into the space of reasons as inferable patterns of discourse, and the associated attribution of intentional states is also taken as part of the inferential articulation to be performed jointly by inferential practitioners. Intentional attribution is here an exercise in discursive practices. The origins of intentionality can therefore only be traced back to the mutual attributions in discursive practices, as Brandom writes, "only communities, not individuals, can be interpreted as having original intentionality. For this to work, the practices that institute the sort of normative status characteristic of intentional states must be *social* practices."[35] By ascribing original intentionality to communities of social practices, intentionality as the mark of mental phenomena is no longer afforded by individuals but by intersubjective practices. To acknowledge that communities can be ascribed original intentionality is to allow their members to attribute intentionality to each other in the sense of mutually authorizing normative statuses. In Brandom's words, "[i]nterpreting a community as exhibiting original intentionality is taking its members to adopt the discursive scorekeeping stance toward each other."[36] This account of intentionality rejects

---

[33] Brentano argues that every mental phenomenon "includes something as object within itself" and can be characterized by "reference to a content, direction toward an object (which is not to be understood here as meaning a thing), or immanent objectivity." Franz Brentano, *Psychology from an Empirical Standpoint, trans. by A.C. R*ancurello, D. B. Terrell, and L. McAlister (London: Routledge, 1995), 68.

[34] Cf. Shaun Gallagher and Katsunori Miyahara, "Neo-Pragmatism and Enactive Intentionality," In *Action, Perception and the Brain,* ed. J. Schulkin (Basingstoke: Palgrave Macmillan, 2012): 118ff.

[35] Brandom, *MIE*, 61.

[36] Ibid., 631.

reducing intentional states to inherent capacities that individuals have by nature. Mutual attribution of intentionality and related representational content is necessary for actualizing one's practice. Put differently, agents endowed with intentionality are the result of rational agency arising from social practices.

This understanding of agent and agency can be compared to the account of relational agency, in which agents are not considered as autonomous individuals, but rather as "interactants" or "interdependents" who are situated in multiple social relations within historical process of institutionalization.[37] According to the relational agency account, normative institutionalized agency does not derive from a single agent, but from the interactions of actors embedded in network of social relations. This definition of agency refers broadly to action and its effect, including habitual activity without explicit realization, and is not limited to autonomous reflective action. This view reverses the relationship between agent and agency.

In the institutionalization of social practices, linguistic practitioners who exercise inferential articulation act as rational agents responsible for their discursive practices. By exercising inferential articulation, the semantics of language use and the intentional states of language users are converted into the norms of how expressions are appropriately used and their normative statuses. Brandom considers that the inferential role of semantic content "can be explained in the pragmatic theory as implicit in discursive practice (which includes intentional agency)".[38] This approach provides an alternative way to articulate the attribution of mental states to linguistic practitioners implicit in discursive practices. He writes:

> In summary, in the theoretical place usually occupied by the notion of *intentional states*, the pragmatics presented here elaborates a conception of *normative statuses*; in the place usually occupied by the notion of *intentional interpretation*, it puts *deontic scorekeeping*-that is, the social practices of attributing and acknowledging commitments and entitlements, which implicitly institute those statuses.[39]

Accordingly, theories of intentional states can be replaced by a normative conceptual framework with entitlement and commitment at its center. The act of intentional attribution and interpretation is considered here as an inferential move in discursive practices. That is, the individual's intentional states and the ability to attribute such states are understood as serving the inferential articulation in discursive practices, and therefore subject to the norms of practice. This normative conceptual framework attributes and tracks not internal states of individuals, but rather intersubjectively shared commitments and entitlements. Understanding others is not a matter of theorizing about their hidden mental states, but rather tracking normative statuses of what they say and do, and predicting what would be reasonable to say and do.[40]

---

[37] Cf. Ian Burkitt, "Relational Agency: Relational Sociology, Agency and Interaction," *European Journal of Social Theory* 19, no. 3 (2016): 322ff.

[38] Ibid., 133.

[39] Ibid., xvii.

[40] Gallagher and Miyahara note that in social cognition the prevailing theory-theory or simulation theory presupposes an "internalist models of intentionality" as representational capacity of the mind. Instead, the account of intentionality from the pragmatist tradition argues that "intentionality

Brandom uses the term deontic scorekeeping to describe this process of tracking the normative statuses of oneself and others in discursive practices, while linguistic practitioners can accordingly be described as deontic scorekeepers. By adopting the deontic attitudes of attribution and acknowledgement, discursive practices become language games of mutual ascription and tracking of normative statuses, as he writes:

> Social practices are games in which each participant exhibits various deontic statuses-that is, commitments and entitlements-and each practically significant performance alters those statuses in some way. The significance of the performance is how it alters the deontic statuses of the practitioners. Looking at the practices a little more closely involves cashing out the talk of deontic statuses by translating it into talk of deontic attitudes. Practitioners take or treat themselves and others as having various commitments and entitlements. They *keep score* on deontic statuses by attributing those statuses to others and undertaking them themselves.[41]

The place where normative statuses occurs is the space of reasons shared by rational agents rather than internal to the individual. Since original intentionality can only be traced back to mutual attribution in communities of linguistic practices, intentional states of individuals that can be explicitly articulated are necessarily dependent on mutual recognition and authorization. This indicates that the normative force of social practices not only constrains individual actions but also, even more importantly, plays a constitutive role in intentional actions.

The constitutive role played by normative forces can guide the intentional actions of human and nonhuman agents. Because fundamentally, the normative inferentialist account of intentionality is in line with a pragmatist understanding of intentionality as a practice of skillfully engaging with environments and objects.[42] Gallagher and Miyahara interpret this understanding as operative intentionality, which is more basic than what is attributed as internal states of mind or body.[43] Operative intentionality is explained as embodied in the process of action and response in movement, "distributed over brain-body-environment", and thus can also be characterized as enactive intentionality.[44] People can practice operative intentionality in adaptation to the environment as well as in social interaction in the form of skilled responses without having to involve representations in the sense of internal states of the mind. Just as in long-term practiced social interaction, people do not have to

---

has a normative status specified by social norms, publicly accessible to other people as well as to the one to which intentionality belongs." Gallagher and Miyahara, "Neo-Pragmatism and Enactive Intentionality," 129.

[41] Brandom, *MIE*, 142.

[42] Brandom remarks: "A founding idea of pragmatism is that the most fundamental kind of intentionality (in the sense of directedness towards objects) is the practical involvement with objects exhibited by a sentient creature dealing skillfully with its world." Robert Brandom, *Between Saying and Doing: Towards an Analytic Pragmatism* (Oxford: Oxford University Press, 2008), 178.

[43] Cf. Gallagher and Miyahara, "Neo-Pragmatism and Enactive Intentionality," 125. They note that the concept of operative (fungieren de) intentionality was introduced by Husserl to describe intentionality involving bodily experience as distinct from intentionality in the sense of mental states (120).

[44] Ibid., 137.

explicitly attribute internal states to others or to themselves, but rather treat perceived actions as affordances that provide possibilities for their own actions, that is, "I see the actions as an affordance for my own possible action (which may be very different from hers); I see the action as inter-actionable or as calling forth a response on my part".[45] What is "seen" here is not an explicit representation, but a recognition of the conditions under which an action is possible and of one's disposition to act. This account takes intentional actions, or more precisely, recognizable patterns of intentional actions, as affordances for the possibility of interpretable further action by an agent. The concept of affordance becomes here central to the interpretation and attribution of intentional states, i.e., patterns that can be recognized as available.

The ability to recognize and respond skillfully to patterns of action can be practiced and mastered by artificial agents in a digitized technological environment. As pointed out at the end of Chap. 3, digitization processes form an active technological environment that provides its users with directions for action on the one hand, the information processing of agents can be distributed across the interconnected cyberspace. The role of the technological environment in guiding the actions of agents is characterized by Verbeek as material intentionality, which depends on and in turn works on the intentionality of human social practices. The operative account of intentionality can essentially explain the fact that material intentionality is not reducible to an inherent property of a particular agent or entity, but rather to a distributed interplay between the environment and the agents. The operative intentionality distributed in the technological environment is driven by vast amounts of data generated by social practices. A series of distributed information processes, from collecting data by sensors, to offline or online data processing, to the generation of interface feedback and instructions, constitute the different states of technological agents. These states are not fixed functional states, but rather are ways of processing information that are constantly being updated in joint practice with human users. In other words, how artificial information processing agents recognize and process patterns of information is shaped by normative forces.

Understanding the constitutive role of social practices on the intentional actions of agents expands the research orientation of intelligent agents. Evans indicates that multi-agent systems in AI research should no longer take the negative constraints of joint practices on the actions of individual agents as its theoretical starting point, but rather value the positive effects that joint practices of multiple agents play on individual intentional actions.[46] In this orientation, AI agents should not be designed with explicit goals and intentions to act as in the past, but rather allow achievable goals and explicit intentions to arise from interactions and joint practices among agents. Accordingly, the development of intelligent agents capable of intentional action should focus on providing them with an environment in which they can

---

[45] Ibid., 138.

[46] Cf. Richard Evans, "Computer Models of Constitutive Social Practices", In *Fundamental Issues of Artificial Intelligence*, ed. Vincent Müller (Cham: Springer, 2016): 389–409.

jointly practice, and in which they can interact to form recognizable information patterns for action. The technological environment, supported by ICTs, the internet, and semantic technologies, is already able to provide such a general environment for networked smart devices. The question of how to assign concrete normative statuses to their operational states is crucial for regulating technological procedures and for responsible participation in discursive practices by technological agents.

To summarize, using a normative conceptual framework to describe intentional attributions and interpretations in discursive practices, the mark of the mental is no longer a particular property of individuals, but a dynamic relationship distributed in interaction with others and environments. Intentionality is understood here in priority as directed operation and then as a mark of the mental. If such ability of directed action is to be modeled, then it should not be treated solely as an internal fixed function of the acting individual. In other words, it is wrong to locate the function of human intelligence inside the skull or body of an individual. There is no ghostly intelligent function residing inside the individual, rather, shared social practices provide the conditions for rational agency to practice mutual attribution of normative statuses. To attribute normative status to others and to oneself is to see others and oneself as deontic scorekeepers who are able to act and interact within reciprocally recognized normative procedures.

5.3 Inferential Articulation of Norm-Making

Brandom's deontic scorekeeping model provides a normative conceptual framework for articulating entitlement and commitment implicit in discursive practices. The core idea is that linguistic practitioners can engage in discursive practices by licensing themselves and others as agents capable of rational action, mutually attributing deontic statuses to each other in the process, keeping deontic scores, and regulating scores through "the production and consumption of reasons".[47] In this framework, the account of normative statuses replaces the representationalist view of mental states. The deontic scorekeepers are knowers and interpreters who maintain the inferential articulation on the one hand, and reliable agents who are responsible for sustaining deontic status on the other. The rest of this chapter will explain the groundwork for this framework, as well as whether and how it can be extended to AI agents that interact with humans.

---

[47] Brandom, *MIE*, 474.

5.3.1 Recognition of Normative Statuses and Reciprocal Authorization

So far, the game of giving and asking for reasons in the space of reasons can be regarded as an extended version of the triadic interaction and triangulation of multiple linguistic practitioners in the shared environment. Since linguistic practitioners can license individual experiences into the space of reasons to be justified, they can put the attribution of causes that lead to their perceptual beliefs into the game of giving and asking for reasons. Linguistic practitioners who engage in inferential games undertake not only the ontological commitment of interpreting common causes, but also the rationalist commitment in discursive practices of authorizing themselves and other linguistic practitioners to make reasonable articulations in the space of reasons. The principle of charity toward others can be thus understood as a rationalist discursive commitment that keeps the discursive practice of justifying knowledge ongoing.[48] And to undertake the rationalist discursive commitment is to license expressions as inferable, to authorize linguistic practitioners as rational agents who can share the space of reasons, i.e., "what I take-true I thereby, ceteris paribus, authorize you to take-true."[49] Such authorization commonly occurs implicitly in discursive practices. Analytically speaking, it consists of key steps that include building discursive patterns with inheritable normative status and mutual authorization among deontic scorekeepers.

The first key step is to establish patterns that allow rationalist commitments to be inherited in shared discursive practices. Specifically, this involves performing assertoric speech acts. Brandom considers that the basic model of inferential practice is to "institute assertional significance", by which commitments and entitlements, the two normative statuses, can be inherited.[50] Commitments and entitlements are defined as two basic terms that describe normative statuses, or rather, deontic statuses of speech acts.[51] These two are incompatible and are inherited in a hybrid way in assertoric speech acts. Assertoric speech acts can provide the basic available

[48] McDowell compares Sellars' logical space of reasons with Davidson's approach to rationality as a constitutive ideal, arguing that both stress the normative constraints on the knowability and interpretability of experience. Davidson takes the principle of charity as the initial attitude of interpretation, allowing the unfamiliar expressions of others to be taken as meaningful, i.e., as true in certain conditions. Sellars takes placing into the logical space of reasons as a necessary move for an episode or state to be understandable, allowing one's own experience to be shared and justified. Both agree that what is observed and the report of it, if any, can only be made understandable to others as well as to oneself by "modeling them on overt linguistic performances", that is, by being subject to the normative constraints of discursive practices. Cf. John McDowell, "Sellars and the Space of Reasons," *Analysis. Claves de Pensamiento Contemporáneo*, (2018): 3f.

[49] Brandom, *MIE*, 239.

[50] Ibid., 168.

[51] Ibid., 160. Brandom argues that commitments and entitlements as basic terms can provide a more general conceptual framework for describing norm-making procedures than the traditional deontological terms of obligation and permission, or duties and rights, which describe norms that have already been prescribed.

pattern for the game of giving and asking for reasons, because making an assertion expresses publicly what the perceptual belief commits to, allowing it to be licensed as a link in the inferential articulation.[52] Assertoric speech acts undertake thereby commitment that the content of beliefs is assertable and judgeable, so that "belief can be modeled on the kind of inferentially articulated commitment that is undertaken or acknowledged by making an assertion".[53] The fact that assertions can inherit deontic statuses is based on the recognition of the law of identity as "*p*, therefore *p*". The recognition of the law of identity allows an assertion to be repeatedly self-cited as a reason. This recognition of self-citation is the basic pragmatic commitment that allows the content committed to by the assertion and its deontic status to be inherited in discursive practices both by the same person and interpersonally.[54]

By making an assertion and assuming such a commitment, one authorizes further inferential moves on the grounds of that assertion on the one hand, and commits to giving reasons for it on the other hand. That is, by performing assertoric speech acts one can institute the deontic statuses of entitlement and commitment that can be altered and inherited in the process of giving and asking for reasons. An assertion in discursive practices is by default entitled, and the inheritance of its entitlement is also by default until it is challenged.[55] Default and challenge form the basic structure of inheriting deontic statuses of assertions. Once an assertion is challenged, it loses its entitled. And as soon as the speaker provides reasons for their assertion, it regains its entitlement. This mechanism of default and challenge alters and tracks the deontic statuses of assertions in the process of giving and asking for reasons.

The other crucial step, occurring simultaneously with the constitution of inheritable normative status, is the reciprocal authorization of linguistic practitioners as potential rational agents. According to Brandom's analysis, demonstrative and personal pronouns can be treated as basic linguistic tools that provide linguistic practitioners with expressions of mutual recognition and reciprocal authorization. By

---

[52] Ibid., 161ff. Using the metaphor of entrance tickets, he outlines how language practitioners assert deontic statuses through performing speech acts. An entrance ticket is an inferential license that allows the practitioner performing the speech act to have a deontic status. In the consumption and use of entrance tickets, the ticket consumer undertakes the commitment to assume the deontic status of performing a speech act, i.e., being responsible for performing a speech act. Brandom notes that the deontic statuses implicit in linguistic practices can be articulated through two corresponding practical deontic attitudes, attribution of others' commitments or entitlements, and recognition of one's own commitments or entitlements.

[53] Ibid., 157.

[54] Brandom distinguishes two different kinds of inheritance, in the first case the same interlocutor makes different assertions, and the commitments they assume inherit the entitlement. This case deals with the "intercontent" and "intrapersonal" inheritance of deontic statuses. The second is the entitlement of different interlocutors to hold commitments to the same assertion and their inheritance. This case involves "interpersonal" and "intracontent" inheritance. Both cases require assertions to be available as patterns that can be self-cited, i.e., to undertake the fundamental pragmatic commitment. Ibid., 175.

[55] Ibid., 177.

using these basic linguistic tools, linguistic practitioners can grasp themselves and each other and their world in terms of a spatial metaphorical image, that is, describe each other as "knowers-and-agents-in-the-world".[56] In this way, linguistic practitioners construct a normative realm of meaning to be shared with others. The referents of place and person are not only knowledge, but also acknowledgment and agreement. To articulate what happens is to recognize it with a normative attitude. It is worth noting that this approach strongly assumes that pre-conceptual episodes of awareness and affect can be identified as cognitive states by means of linguistic tools.[57] Although the exact articulation varies according to the habitus of the linguistic practitioner, there is a general normative stance on the use of spatial and personal references that can be shared with others.[58]

To begin with, the use of demonstrative pronouns, especially when using them to denote something, is a key move for linguistic practitioners to begin to share their own states.[59] Typically, demonstrative pronouns such as "this" and "that" can be learned by combining them with indicative gestures or postures. Once mastered, these terms can be used as linguistic units to express referents.[60] The use of demonstrative pronouns acquired in this way has a substitutional function, i.e., they can be used to refer to and substitute for events that are jointly committed as known in shared practice.[61] The substitution function allows demonstrative pronouns to serve as abstract expressions and refer to different perceptual beliefs. To put it more specifically, the use of demonstrative pronouns enables the irreducible concrete perceptual "tokens" associated with demonstrative acts to be replaced by repeatable "types", allowing individual experiences to become expressions that can be communicated in discursive practices.[62] In the repeatable use of demonstrative pronouns,

---

[56] Brandom, *Brandom reads Sellars*, 13.

[57] Dreyfus and McDowell had a debate about whether intelligent activities prior to conceptual expression were nonconceptual or conceptual. McDowell argued that such activities involved conceptual capacities and were therefore not nonconceptual, while Dreyfus and Taylor later adopted the term preconceptual. Hubert Dreyfus and Charles Taylor, *Retrieving Realism* (Cambridge, MA: Harvard University Press, 2015), 77ff.

[58] Ibid., 116ff. Dreyfus and Taylor use the term habitus, coined by Bourdieu, to illustrate how agents already live with others in a space of shared social meanings before an explicit access to the space of reasons is available.

[59] Brandom, *MIE*, 462. Recent studies on early childhood development also suggest that demonstrative pronouns are among the earliest and most frequent linguistic utterances made by young children that do not derive from other word roots. This suggests that learning how to use spatial demonstrative pronouns in a normative way is more primitive than mastering proper names referring to specific objects and persons. Frauke Hildebrandt, Ramiro Glauer and Gregor Kachel, "Coming from a world without objects," *Mind & Language* 37, no. 2 (2022): 161.

[60] Brandom, *MIE*, 479.

[61] Ibid., 462ff.

[62] Habermas indicates that: "The intralinguistic recurrence to singular demonstrative acts replaced by demonstrative pronouns was supposed to explain how unrepeatable 'tokens' – link-ups with perceived events – can be brought into the discursive chain of repeatable 'types', thereby being made recursively accessible." Jürgen Habermas, "From Kant to Hegel: On Robert Brandom's Pragmatic Philosophy of Language," *European Journal of Philosophy* 8 (2000): 338.

what is perceived to trigger perceptual beliefs is substituted and allowed to be constantly repeated and recurred. Notably, the initial substitution of demonstrative pronouns for perceptual beliefs is different from the later repeatable use of them. Initial substitution is the substitution of tokens with types, whereas the repeatable use after permitting substitution is called anaphora.[63] By using demonstrative pronouns, the anaphoric chain is initiated, in which the use of pronouns maintains the committed referential relationship between repeatable linguistic types and original perceptual tokens.

Then, Brandom remarks that the use of the personal pronouns "I" and "you" is similar to the use of demonstrative pronouns "this" and "that". He writes that "[i]n the same way, tokenings of 'you' could be understood as anaphorically dependent on (perhaps merely possible) tokenings of 'I' by others."[64] By using the pronouns "I" and "you" interchangeably to refer to oneself and others in discursive practices, linguistic practitioners are substituted by an "I-Thou" anaphorical relation that facilitates the interpersonal communicative inheritance of deontic statuses. To put it differently, the use of first- and second-person pronouns is an important step in licensing of linguistic practitioners into the normative space of reasons, allowing them to become rational agents who can entitle and undertake commitments. Thus, starting from the basic "I-Thou" relation, a set of general social relations about the "self-other" can be constructed.[65] On the one hand, the expressions of subjective attitudes in interaction are denoted by the interchangeable use of personal pronouns "I" and "you". By using them linguistic practitioners place themselves in shared discursive interaction with others as objects that can be attributed states. The interchangeability of "I-Thou" allows the understanding of the content conveyed by an expression not to be limited by the authority of the first person, but to be inherited interpersonally as a commitment to the content in discursive practices. On the other hand, in using the "I-Thou" structure, a language practitioner tacitly allows other interactants to use it as well, so that "I" and "you" make sense as pronouns. In other words, using "I" and "you" as personal pronouns provides symmetrical rules of

[63] Anaphora is understood as "a special mechanism for the inheritance of substitution-inferential commitments". Brandom, *MIE*, 283. The recurrence of a demonstrative pronoun that bears a commitment to the same initial substitution implies the inheritance of its deontic status. Thus, the use of demonstrative pronouns can be regarded as "a special kind of anaphoric initiator" (Ibid., 455).

[64] Ibid., 463.

[65] The correlation between language acquisition and the cognitive ability to shift perspectives has been demonstrated in developmental psychology research. Fuchs notes that in the transition from dyadic to triadic interactions, children's comprehension of perspective shifting can develop from an "I-Thou" to a "Self-Other" meta-perspective, which is understood as "explicit awareness of self-other interaction" and "may include perspective-taking (explicit third-person perspective) as well as self-reflection (explicit first-person perspective)". With the full development of perspective shifting skills that accompany linguistic competence, it is possible to build explicitly recursive sentences, such as "I think that you think what I think", expressing the mutual recognition as subject and object of mental propositional content. Cf. Thomas Fuchs, "The phenomenology and development of social perspectives", *Phenomenology and the Cognitive Sciences* 12, no. 4 (2012): 660ff.

language use for oneself and others. Linguistic practitioners employing them undertake a shared commitment to their use. Consequentially, the "we" discursive community is formed in the way that linguistic practitioners adopt such a practical attitude towards each other, i.e., recognize each other and allow each other to recognize oneself as an object that can be referred to by the "I-Thou" structure.[66]

Taken together, with the help of the referential frame provided by linguistic structures, linguistic practitioners can identify each other and license each other as rational agents sharing a normative space. In other words, practitioners can mutually recognize and reciprocally authorize each other as subjects of deontic statuses, as Brandom writes, "recognition is a *normative* attitude. To recognize someone is to take her to be the subject of normative statuses, that is, of commitments and entitlements, as capable of undertaking responsibilities and exercising authority."[67] This view can be traced back to Hegel's account of self-consciousness as arising from the practice of mutual recognition.[68] In a state of mutual recognition, self-reference is the state of recognizing the other while allowing the other to recognize oneself, where each self-referential action depends on the same move of the other. Hegel considers such state as the condition of self-consciousness, he writes, "self-consciousness exists in and for itself when, and by the fact that, it so exists for another; that is, it exists only in being acknowledged".[69] Mutual recognition is thus

---

[66] Early childhood development research shows that the use of personal pronouns in caregiver-child interactions plays an important role in young children's self-development and social cognition. In using "you" to address young children, caregivers acknowledge the potential of young children as members of the community, which scaffolds young children's access to the shared normative space. The use of personal pronouns conveys attitudes of recognition and expectation. As Musholt points out, understanding oneself and others as "persons" is a normative stance that develops in young children's interactions with their caregivers. This development is a process from implicit awareness at the bodily level and affective response to explicit recognizing upon acquisition of conceptual language, and an essential part of it is "being addressed by others as 'you' and learning to play the social game of giving and taking reasons". Cf. Kristina Musholt, "Self and Others," *Interdisciplinary Science Reviews* 43, no.2 (2018): 143. Also see David Bakhurst, *The Formation of Reason* (Oxford: Wiley-Blackwell: 2011), 52ff.

[67] Robert Brandom, "The Structure of Desire and Recognition: Self-Consciousness and Self-Constitution," *Philosophy & Social Criticism* 33 (2007): 134.

[68] Brandom notes that Hegel lays the foundation for the formation of self-consciousness by articulating the transition from desire to recognition in action through the master-slave dialectic. He outlines the necessary transition in the constitution of the self-conception from an understanding of desire in naturalistic terms to an understanding of recognition in normative terms. Self-consciousness is here understood as a state of being both for the other and thereby for oneself. Cf. Brandom, "The Structure of Desire and Recognition: Self-Consciousness and Self-Constitution", 130 ff. A detailed argument of how the transition of desire to recognition is the basis for intentional action can also be seen in Stekeler-Weithofer's reading of Hegel. See Pirmin Stekeler-Weithofer, "Wer ist der Herr, wer ist der Knecht? Der Kampf zwischen Denken und Handeln als Grundform jedes Selbstbewußtseins," in *Hegels Phänomenologie des Geistes. Ein kooperativer Kommentar zu einem Schlüsselwerk der Moderne*, eds. Klaus Vieweg and Wolfgang Welsch (Frankfurt am Main: Suhrkamp Verlag, 2008), 205ff.

[69] Brandom, "The Structure of Desire and Recognition: Self-Consciousness and Self-Constitution", 134. For the original passage Cf. George W. F. Hegel, *Phenomenology of Spirit*, trans. Arnold

the condition of reflexive recognition, knowing the self by recognizing the other while allowing oneself to be recognized as well. Such knowing is referenced and identified in a normative frame. The discursive practice mode of mutual recognition not only allows the self-referential move to take place, but also authorizes it to take on normative statuses and thus constructs a normative domain. This normative domain is what Hegel calls "spirit (Geist)", where reciprocally authorized agents can act rationally. He writes:

> A self-consciousness exists *for* a *self-consciousness*. Only so is it in fact self-consciousness; for only in this way does the unity of itself in its otherness become explicit for it. [...] A self-consciousness, in being an object, is just as much 'I' as 'object'. With this, we already have before us the Notion of Spirit. What still lies ahead for consciousness is the experience of what Spirit is – this absolute substance which is the unity of the different independent self-consciousnesses which, in their opposition, enjoy perfect freedom and independence: 'I' that is 'We' and 'We' that is 'I'.[70]

This citation indicates that spirit is a normative description of mental states in a collective as well as an individual sense. In its basic form, the "self" that undertakes normative statuses is the self-referential move in discursive practices. The "self" here does not yet include a unified psychological personality or social identity, but rather a minimally sharable form that takes a logical place. This basic form prepares for accountable interactions between rational agents. While "I" can be licensed as an object, it also permits "you" to become the subject and rational agent of discursive practices. The reciprocal authorization enables the "self" to become the subject of entitlement and commitment.[71] The reciprocal authorization as subjects with normative statuses is thus a process of moving "from a world of *desire* to a world of *commitments*, *authority* and *responsibility*".[72] This process is the move from the natural world into the normative domain of the spirit, from the space of causes into the space of reasons. This is an important reception of Hegel's philosophy by the normative inferentialism. In this line, undertaking normative statuses in the community of practice does not restrict the possibilities of individual activity; rather, the shared discursive practice transforms commitments to material inferences in

V. Miller (Oxford: Clarendon Press, 1977), 111 [§178].

[70] Hegel, *Phenomenology*, 110f. [§177]

[71] In Brandom's words: "To be a self, [...] is to be the subject of normative statuses that refer to commitments; it means to be able to take a normative stand on things, to commit oneself and undertake responsibilities." Brandom, "The Structure of Desire and Recognition: Self-Consciousness and Self-Constitution," 125.

[72] Ibid., 133. Brandom stresses that the process of the emergence of normative domains in nature needs to be understood in its formal and real historical context. He writes: "[t] his corresponds to the shift from consideration of particular merely biological creatures impelled wholly by *natural impulses*, in relation to their species, on the one hand, to consideration of genuinely social self-conscious individuals motivated by normative relations of *authority* and *responsibility* within their communities, on the other. How one understands the relation between these, both conceptually and historically, is evidently of the first importance in understanding what Hegel has to teach us about the normative realm he calls '*Geist*'" (134f).

practice into normative forces, extending the sphere of individual practice into the normative logical space of reasons, the normative domain of spirit.

The advantages of mutual recognition as a starting point for attributing mental states can thereby also be viewed from a social cognitive perspective. Mutual recognition and reciprocal authorization through the practice of using symmetrical "I-Thou" structure facilitates the acquisition and defense of empirical knowledge. For by practicing how to refer to self and others and their discursive acts, linguistic practitioners license themselves and others to be rational knowers as well as accountable agents. They can exercise the ability to speak rationally, including using language for attribution, and to act rationally, including adopting practical attitudes. In tracking each other's normative statuses, linguistic practitioners develop the social skills of mutual attribution. Brandom writes:

> This is […] an *I-thou* sociality rather than an *I-we* sociality. Its basic building block is the relation between an audience that is attributing commitments and thereby keeping score and a speaker who is undertaking commitments, on whom score is being kept. The notion of a discursive *community* – a we – is to be built up out of these *communicating* components.[73]

The "I-we" relationship between a single linguistic practitioner and the community presupposes the "I-Thou" relationship, as the exchangeable use of the "I-Thou" structure ensures the interdependence of self and other in discursive practice. Notably, the "I-Thou" relation involves interactors in shared discursive practice, whereas the "self-other" relation as further authorization of normative status involves attributors and those who are attributed. In general, in the normative inferentialist account, the "self" is available in its basic form of a self-referential move in discursive practice in order to be conferred normative status. The self is the placeholder of the normative space established in discursive practices, entitled and committed to normative statuses.[74] Through mutual recognition and authorization, linguistic practitioners can attribute and assess each other's the normative status. This view emphasizes the constructive role of social practices based on discursive interaction for rational agents.

In recent years, mutual recognition has received attention in fields such as human-robot interaction and social robotics. The main concern is how recognition can underlie mutually adaptive interactions, and consequently the social interaction skills of robots.[75] Brinck and Balkenius point out that moving from an instrumental to a normative stance toward robots facilitates the development of human-robot interactions.[76] For human users, recognizing artifacts such as social robots as

[73] Brandom, *MIE*, 508.

[74] Robert Brandom, *Tales of the Mighty Dead. Historical Essays in the Metaphysics of Intentionality* (Cambridge, MA: Harvard University Press, 2002), 16. Brandom indicates that Kant makes the normative turn from the ontological description of "selves as thinking beings, to a deontological demarcation of selves as loci of responsibility" (16).

[75] Cf. Ingar Brinck and Christian Balkenius, "Mutual Recognition in Human-Robot Interaction: a Deflationary Account," *Philos. Technol.* 33 (2020): 53.

[76] Ibid., 64.

collaborators means treating them not only as tools to accomplish tasks, but also as agents who can be held accountable for their actions. They argue that through such recognition a "we-space" can be generated in which mutually adaptive interaction can take place between reciprocally recognized interactors.[77] In this "we-space", the actions of interactants couple dynamically in a reciprocally coordinated manner. For instance, the process of two individuals moving an object together requires that the interaction partners be recognized as collaborators. This line of thought is consistent with the normative turn of mind, where the generation of "we-space" is the mutual recognition and authorization into normative space.

It is important to note that considering artifacts as collaborators does not mean treating them as human-like, but rather recognizing their role as collaborators in interactive processes such as gaze, sound, touch, and other specific sensorimotor-based processes. Brinck and Balkenius indicate that deliberately misleading human users to recognize emotionless robots as persons through design tricks could instead raise ethical issues.[78] Mutual recognition defined in a deflationary manner is therefore more helpful in guiding human-robot interactions supported by current robotics.[79] This definition of recognition is in line with the original meaning of recognition as emphasized in normative inferentialism, namely, its meaning as information pattern recognition rather than acknowledgement. Recognition is here the basic modality that allows something to be identifiable so that it can be licensed into discursive practices. Such recognition is more initial than the acknowledgement of personhood and personal identity. The word "person", etymologically, means the character wearing a mask in ancient Greek drama, playing a certain social identity while performing related actions, and is later extended to a person with legal rights.[80] For human agents, the recognition of personal identity is integrated into the basic recognition of rational agents in interaction. But for technological agents, mutual recognition can currently only be restricted to the level of rational agents for the sake of avoiding ethical issues. The more predictable the action patterns of rational agents are in accordance with certain character traits, the closer they are to the manifest image of man in the world based on the concept of person.

In general, the normative inferentialist account demonstrates that normative statuses can only be authorized and challenged intersubjectively. In shared practices, the "self" of linguistic practitioners is legitimized as the subject who can attribute

[77] Ibid., 58. They explain that "[w]e-space gives the agents access to social affordances as well as social norms and values built into artifacts and material infrastructure, with new forms of interactive behavior as the result. The emergence of we-space becomes apparent in situations where the space clearly is created by the interaction."

[78] Ibid., 59ff. They point out that some commercial robots are designed to imitate basic social competences, such as responding to human emotional expressions, but without further reciprocal interaction capabilities with users. Such designs evoke the emotions and expectations of human users, exploiting the emotional attachment that may arise to make users dependent on the products, which is arguably unethical.

[79] Ibid., 67.

[80] Jenny Techman, "The Definition of Person," *Philosophy* 60, no. 232 (1985): 177.

and be attributed with mental states, authorized as the responsible rational agent who can license and be licensed with normative statuses. Mutually licensed and attributed linguistic practitioners afford a wider range of practical possibilities to each other. Taking on the rationalist commitment to discursive responsibility, linguistic practitioners place both perception and action into the logical space of reasons. This suggest that for responsible agents in discursive practices, whether human or artificial, to be recognized as rationally responsive, they need to be licensed to have access to normative status, to be allowed to practice inferential articulation in the game of giving and asking for reasons.

5.3.2 Expressing Practical Reasoning with "Pragmatic Metavocabularies"

In the normative inferentialist approach, the rational agency of discursive practices constantly exercises inferential articulation. Such exercises not only provide appropriate reasons for linguistic expressions, but also allow linguistic practitioners to keep normative statuses. In addition to establishing discursive patterns and conferring normative statuses on linguistic practitioners, the discursive responsibility for maintaining inferential articulation is also crucial for the game of giving and asking for reasons. Responsibility is understood here as the responsiveness of linguistic practitioners in their discursive practices. This is a general account of how authorized practitioners can be held accountable and expected to be responsible for discursive practices in a shared practice. Although norms may differ widely across different practice communities, the enforcement of and expressions of norms can be generally characterized as maintaining of discursive responsiveness.

Discursive responsiveness is the responsibility to mutually track and justify normative status, which is also called "justificatory responsibility" by Brandom.[81] The close relationship between discursive responsiveness and justifying normative status can be regarded in two aspects. On the one hand, normative statuses are instituted and inherited by discursive practitioners through performing assertoric speech acts. Making assertions and then giving or asking for reasons are the basic moves that institute, inherit and challenge their normative statuses. On the other hand, linguistic practitioners can show that they are discursively responsive by making assertions. By maintaining responsiveness in discursive practices, linguistic practitioners can justify themselves as subjects of normative statuses.[82] By undertaking the

---

[81] Brandom, *MIE*, 173.

[82] The close relationship between responsibility and responsiveness is also highlighted in Derrida's reading of Heidegger's notion that "speaking is responding". In this view, "'response-ability' marks the idea that every action, discursive or otherwise, is only born of an engagement with the set of conditions that produced it. This is the logical implication of the idea that every act of speaking is an act of response." Christian Lundberg & Joshua Gunn, "Ouija Board, Are There Any

justificatory responsibility, linguistic practitioners can "show that they are entitled to the commitment expressed by their assertions".[83]

The fulfillment of the justificatory responsibility is conditioned on assertions being challenged and queried. Such responsibility can thus also be understood as "a conditional task responsibility" that linguistic practitioners take on when making assertions.[84] More specifically, when an assertion is challenged and queried, the assertion loses its default inherited entitlement. For the assertion to regain its committed normative status, the linguistic practitioner is obliged to justify the assertion by giving reasons for it. Linguistic practitioners are responsible for challenging and defending the default inherited entitlement and commitment of assertions. The way to challenge and justify is to keep putting forward new assertions. The essential idea is that an assertion can be viewed as a responsive action that "occupies a position in an inferentially articulated space of claims that can be offered as, and stand in need of, reasons".[85] Thus, the justificatory responsibility is a response to do something, to make reasonable inferences in the space of reasons.

Brandom emphasizes that making assertions as the basic step in inferential articulation is both "saying" and "doing".[86] In discursive practices, performing assertoric speech acts takes on not only doxastic commitments, but also practical commitments, which are two types of discursive commitments. Undertaking a doxastic commitment is taking the content of an assertion to be true, while undertaking a practical commitment is regarding making an assertion as an intentional act. To attribute an act as intentional means to entitle it to undertake a practical commitment, and this entitlement brings the act into the game of deontic scorekeeping. Accordingly, it is the practical commitment that allows scorekeepers to attribute and elaborate on each other's intentions to act. Doxastic and practical commitments provide complementary responsiveness in discursive practices for practitioners performing practical reasoning. Taking what is said to be true as well as treating what is done to be intentional are, in effect, licensing discursive commitments, which are mutually licensed, challenged, and inherited by language practitioners in shared practices. The skillful licensors and takers of discursive commitments are rational agents. In Brandom's words, "we are rational creatures exactly insofar as our acknowledgment of discursive commitments (both doxastic and practical) makes a difference to what we go on to *do*."[87] Acknowledging discursive commitments requires that discursive practitioners authorize each other as deontic scorekeepers who can mutually attribute and track deontic statuses.

Communications?" Agency, Ontology, and the Death of the Humanist Subject, or, Continuing the ARS Conversation, *Rhetoric Society Quarterly* 35, no. 4 (2005): 97.

[83] Brandom, *MIE*, 173.

[84] Ibid., 178.

[85] Ibid., 618.

[86] Ibid., 173.

[87] Robert Brandom, "Action, Norms, and Practical Reasoning", In *Philosophical Perspectives*, 12, Language, Mind, and Ontology (1998): 137.

The responsiveness of rational agents undertakes discursive commitments and is thus no longer a capacity for immediate reflection on stimuli. This is because both "saying" and "doing" in discursive practices are linguistically structured acts that are not only in the chain of causal explanation but also in the articulation of reasons.[88] To put it in another way, rational agents that carry discursive commitments are located both in the material space of causes and in the logical space of reasons. The expressive role of logic in a broad sense allows the space of causes to be rendered in discursive practices by the logical space of reasons. That is, the material practical procedure can be expressed by law-like rules in inferential articulation. The possible patterns of practice and structures of the world can be expressed in terms of inferential relations.[89] Inferentially expressed practical procedures are not descriptive, but carry doxastic commitments and practical commitments.

Accordingly, "the capacity to respond reliably to acknowledgements of commitments" is understood as practical reasoning.[90] The concept of the law plays an essential role in rationalist philosophy since Kant, where practical reason is understood as "the capacity to derive performances from a conception of laws".[91] That is, practical reason is the capacity to be willing to act driven by certain laws. The debate over what kind of laws could generate appropriate actions, or how the actions could be derived from the laws, has been an ongoing controversy. Brandom argues that addressing this problem requires a pragmatic version of how to derive performances from laws, that is, an account of how laws are enacted and used as rules. He describes the relationship between the binding force of law and performances in linguistic practice in normative terms as follows:

> 'Law' is Kant's term for a binding rule—a norm. One's conception of a law is what one takes oneself to be obliged to do. Having a rational will, then, can be understood as having the capacity to respond reliably to one's acknowledgment of a commitment (of a norm as

[88] More specifically, Brandom distinguishes three kinds of causal relations between action and reason: acting intentionally, having reasons for action, and acting for reasons. "1. an agent's acting intentionally-that is, acknowledging a practical commitment by producing a performance or exercising a reliable noninferential differential disposition to respond to acknowledgments of practical commitments by producing a performance, 2. an agent's having reasons for action or acting with reasons-that is being entitled to a practical commitment, and 3. an agent's acting for reasons, the action being caused by (attitudes toward) reasons for action-that is the acknowledgment of the practical commitment having arisen by a process of inference from acknowledgment of the commitments that provide the entitling reasons." (Ibid., 136; *MIE*, 262) These are three typical acts associated with the articulation of discursive commitments and actions. Acts can be attributed to reasonable actions by bringing them into the game of giving and asking for reasons. That is, by treating them as responses that can be formed as assertoric speech acts so that they can be placed into the inferentially articulated space of reasons.

[89] Redding sums up this idea concisely as "the structure of 'being' is that which *shows itself* within the logical structure of our *sayings about* particular understanding logical here as comprising those features of our sayings that mediate their inferential relations." Redding, "Analytic Philosophy and the Return of Hegelian Thought," 232.

[90] Ibid., 270.

[91] Ibid.

binding on one) by differentially producing performances corresponding to the content of the commitment acknowledged.[92]

Kant's conception of law characterizes the self-legislation of social practice in terms of mutual authorization and binding, which means that practitioners attribute to each other the authority and responsibility of legislators. Accordingly, the capacity for practical reasoning is treated as a universal capacity based on self-legislation, that is, the will of a robust rational agent. This capacity makes lawmaking a universal procedure, although the content of commitments in normative regulations can be diverse. That is, the most basic capacity of a rational agent is not to comply with given fixed rules, but to be able to license different normative contents, to assume commitments to them, and to respond to acknowledgments of them in a reliable way. Such capacity is analogous to the capacity to make dispositional commitments in discursive practices that can be articulated by linguistic structures like modal terms.

To provide a more detailed account, Brandom examines the use of normative vocabularies such as "should", "shall", and "ought" in practical reasoning. He analyzes three typical patterns of practical reasoning: the prudential(instrumental) "ought", the institutional "ought", and the unconditional "ought", which can also be called prudential reasons, conventional reasons, and moral reasons.[93] Three corresponding examples are given below:

(α) Only opening my umbrella will keep me dry, so I shall open my umbrella.
(β) I am a bank employee going to work, so I shall wear a necktie.
(γ) Repeating the gossip would harm someone, to no purpose, so I shall not repeat the gossip.[94]

These patterns of reasoning can be formulated by sentences with modal verb "shall" to indicate an acknowledgement of a practical reasoning.[95] This acknowledgement clearly makes a practical commitment for further action of the linguistic practitioner. So how does such endorsement occur, is it derived from some law of formal reasoning? Brandom notes that if we take these practical reasonings as part of formal inference, then there are some implicit premises that need to be stated to draw the above conclusions. He adds the following premises to each of these reasonings:

(a) I want (desire, prefer) to stay dry.
(b) Bank employees are obliged (required) to wear neckties.

[92] Brandom, "Action, Norms, and Practical Reasoning", 137.

[93] Brandom, *MIE*, 252. These three are representative models of practical reasoning, but they do not exhaust all types of practical reasoning.

[94] Ibid., 245.

[95] Remarkably, Brandom compares the similarities and differences in the use of "shall" and "should". "Should" is typically used when one expresses a practical reasoning from a third-person perspective, such as "He is a bank employee, so he should wear a necktie". In contrast, "shall" is usually used in the first-person singular and plural. He argues that "should" is an attribution and assessment of others' practical commitments, while "shall" is a deliberative attitude that internalizes that assessment into the first-person perspective. 137f.

(c) It is wrong (one ought not) to harm anyone to no purpose.[96]

Combining (a)(b)(c) with (α)(β)(γ) can form a formally complete form of reasoning. For instance, Premise (1): I want (desire, prefer) to stay dry; Premise (2): Only opening my umbrella will keep me dry; Conclusion: I shall open my umbrella. This treatment appears to form a complete syllogism. But one does not usually need to make complete formal inference when making practical commitments and acknowledging them. One may also give other additional premises to change the practical reasoning. In fact, action usually precedes explicitly expressed reasoning in practical reasoning. One acts first and then gives the most likely explanations for the action. I may have opened my umbrella first and then attributed to myself the desire to stay dry, while knowing that the only way to stay dry is to open my umbrella. This is an attribution of the intention to act after it has occurred, and the attribution indicates an endorsement of this practical reasoning. Brandom points out that practical reasoning is usually non-monotonic, which means that they are defeasible reasonings that are allowed to be withdrawn when challenged and questioned.[97] This feature suggests that practical reasoning takes on dispositional commitments that can be expressed as law-like inferential patterns through linguistic structures like modal terms.

Deriving conclusions in practical reasoning and using descriptive expressions in material inference are both inference patterns that imply commitments. For example, "It is raining, *therefore* the street will be wet" and "It is raining, I *shall* open my umbrella" have similar patterns of inference, because both imply commitments that can be expressed, whether to material inferences or to beliefs.[98] Just as the expression of material inference implies acknowledgement of commitments to material rules of inference, the expression of practical reasoning can be understood as acknowledgement of implied practical commitments. We can use modal or normative vocabularies to make explicit the practical commitments that are implicit in practical reasoning, just as we use logical tools to express commitments to material inferences. Accordingly, the term "shall" in the sentence "it is raining, I *shall* open my umbrella" plays the similar role as the term "therefore" in the sentence "it is raining, *therefore* the street will be wet", expressing a licensed law of derivation that can be challenged. This brings us back to the expressive role of linguistic structures, especially logical terms in a general sense. Using specific expressions to articulate practical reasoning is not to formally complete the premises of how an action occurs,

---

[96] Brandom, *MIE*, 246.

[97] Brandom, "Action, Norms, and Practical Reasoning", 132f. He defines monotonic reasoning as "if the fact that the inference from p to q is a good one meant that the inference from $p\&r$ to q must be a good one". (132) That is, further evidence does not affect a given good inference. In contrast, non-monotonic reasoning is inference that can be retracted or devised based on new evidence, which includes default logic, abductive reasoning, and belief revision.

[98] Brandom compares the role of normative terms and conditionals in expression: "Normative vocabulary (including expressions of preference) makes explicit the endorsement (attributed or acknowledged) of *material* properties of *practical reasoning*. Normative vocabulary plays the same expressive role on the practical side that conditionals do on the theoretical side." Ibid., 134.

but to express how it is permitted to occur. As Brandom puts it, "[t]he idea is that normative vocabulary is a kind of logical vocabulary, in my expressive sense: its expressive function is to make explicit commitments to inferences."[99] Expressions with such linguistic tools can make implicit practical commitments explicit in the form of attributed intentions.

Modal, normative, or evaluative vocabularies that have a similar expressive role to typical logical vocabularies are also termed by Brandom as "pragmatic metalinguistic vocabularies", or briefly, "pragmatic metavocabularies".[100] Expressed practical reasoning helps to clarify implied practical commitments and account for them. Endorsement of practical commitments typically arises from the norms enacted by practitioners and their discursive practice communities. Given that rational agents are reciprocally recognized in shared practices, before they can describe and explain accurately what they do and say, they have already acquired certain patterns of behavior, which have been shaped by norms in their communities. Modals and other pragmatic metavocabularies help to specify the framework within which the language use refers to and commits to it. In other words, these vocabularies can augment the expressive power of language.

The normative force that in effect guides or enforces actions can be expressed in normative or evaluative terms, as in the added premise formulation in above examples. The effect of using these pragmatic metavocabularies suggests that the law-like pattern of practical inference is such a process that licenses itself to carry out inferential articulation and can be challenged, modified, or withdrawn. Brandom writes:

> The idea is that the broadly normative or evaluative vocabulary used in (a), (b), and (c) ('prefer', 'obliged', and 'ought') [...] is used to make explicit in assertible, propositional form the endorsement of a *pattern* of material practical inferences. Different patterns of inference should be understood as corresponding to different sorts of norms.[101]

That is, sentences (a)(b)(c) serve in effect to express in normative terms the linguistic practitioner's endorsement of different practical commitments. Whether one should open an umbrella depends on the endorsement of personal desires or preferences for staying dry. In this prudential(instrumental) practical reasoning, the endorsement of a desire or preference is like taking the content of a doxastic commitment to be true. In the case of institutional "ought", whether one wears a necktie depends on the endorsement of the norm that bank employees are obliged to wear neckties, and whether one possesses the social role of bank employee.[102] And the case of unconditional "ought" involves some universally accepted commitments,

[99] Ibid.

[100] Brandom, *Brandom reads Sellars*, 8.

[101] Brandom, "Action, Norms, and Practical Reasoning", 134.

[102] Ibid., 135.

such as "it is wrong (one ought not) to harm anyone to no purpose", without regard to personal desire or social relations.[103]

With these explicitly expressed stances of practical commitments, it is easier to attribute the actions of practitioners and to trace their deontic statuses in discursive practices. For these vocabularies allow practical reasoning to be challenged and questioned in the same way as doxastic reasoning. The agents who make practical commitments therefore also carry the justificatory responsibility. That is, attributing an action with intention requires clarifying its practical commitment and providing reasons for it.

What rational agents are capable of is thus the ability to perform appropriate speech acts and to be able to give reasons and assessments for doing so. Modal terms and other pragmatic metavocabularies help to articulate the implicit inference patterns upon which actions are performed. The ability to perform such linguistic practices is also termed by Brandom as the "algorithmic ability" of elaborating inferential patterns and the ability to "responsive substitution" of them with ordinary vocabularies.[104] The exercise of this ability has a constructive role to play in shaping rational agents, making them responsive and interpretable in shared discursive practices. An essential implication of the normative inferentialist approach, accordingly, is that human rationality is no longer an innate characteristic of individuals. Rational agents and their reliable ways of acting are reciprocally authorized and enacted in shared discursive practices. Linguistic practitioners institute normative domains through mutual recognition and assuming normative statuses in their shared practices. The experience of action and interaction can be expressed and enacted as law-like norms in reciprocally authorized shared practices. Humans are rational beings not because they comply with some inherent given rules, but rather they articulate the possible rules in shared practices and acknowledge their entitled normative force on individual practices.

To conclude, pragmatic metavocabularies provide deontic scorekeepers with the tools to perform discursive responsiveness and justify normative statuses. The use of these terms facilitates expressing doxastic and practical commitments of speech acts, attributing linguistic expressions as meaningful and acts as intended. The practical reasoning that is expressed can have various patterns. The different kinds of practical reasoning do not need to be generalized into a single form. Since they come from the different practical norms of communities and the practical experiences of individuals, they can be expressed in various terms as different law-like "oughts".[105] What is important is that discursive practitioners, whether from different discursive communities or with widely varying individual experiences, are able to track each other's discursive commitments with the help of discursive metalinguistic vocabularies. In the current development of large-scale language models

[103] Brandom notes that "For Kant the expressive role of 'ought' or 'should' (Sollen) is to make norms explicit in the form of imperatives". Brandom, *MIE*, 270.

[104] Brandom, *Brandom reads Sellars*, 172.

[105] Brandom, "Action, Norms, and Practical Reasoning", 136.

represented by ChatGPT, taking advantage of the expressive power of pragmatic metavocabularies can allow artificial agents to better engage in discursive practices of diverse communities responsively. I will return to this point in Chap. 6.

Indeed, networked artificial agents in smart environments are already performing tasks involving normative statuses in interaction with human or other artificial agents. Social practices have extended the space of reasons into linked cyberspace that provide available resources and possibilities of action for human and artificial agents. The formation and development of the internet is a digital simulation and extension of common discursive practices. Semantic technologies in general refer to social processes of encoding meaning, including both the traditional ways of recording and representing social practices through symbols as well as the current information technologies that enable machines to read data. Expressions with self-referential forms can be recursively used in discursive practices by allowing themselves and others to be in exchangeable positions. Such expressions can be used as both the subject and predicate or object of sentences, for the self-referential move allows expressions to be used as both descriptions and as meta-linguistic descriptions. In this way, the self-referential form and its recursive use facilitate semantic technologies, because the acquired information patterns are allowed to be used as sentence constituents to form clauses and sentence clusters. Information patterns that commit to content can be placed in semi-structured syntactic forms that can be used to describe and be described. This is the guiding principle of the semantic triple model in linked data which allows the identified web resources to be any sub sentential structure of the subject-predicate-object syntactic form. The resources can therefore be interlinked and be described in relation to each other. With this model, the description and definition of resources is not derived from fixed ontologies; instead, resources are allowed to describe each other as metadata with discursive commitments. Whether the description is reasonable or not will be further described in the use of the resources.

The way in which semantic technologies regulate semantic relations in use corroborates the normative inferentialist approach, i.e., licensing and examining semantics from the discursive practices of language communities. With the basic form of licensing semantic relations, practitioners of discursive practices can mutually recognize each other's normative status in the logical space. That is, semantic technologies support the internet in a way that enables networked devices to place their data integration and information processing into a normative conceptual framework. Just as image or sound data recognized by digital devices can be marked and structured with semantic technologies so that they can be interrelated and interpreted in relation to other web resources. These structured resources in turn serve to afford the agents with the basic information patterns for inferential articulation and making meanings. The process of meaning-making is thus in the normative space of social practices, rather than within agents that are presupposed to be isolated from the external world.

With this understanding, misguided inquiries into the mind such as the search for the "ghost in the machine" can then be shifted to the recognition and authorization of rational agents in shared normative practices. Contemporary cyberspace is the logical space for norm-setting and meaning-making in the digitalization of social practices, which can provide shareable resources and discursive practice environments for both human and artificial agents. The question that remains is how to enable agents engaged in discursive practices in cyberspace to take a normative stance to authorize others and to allow themselves to be authorized. That is, along these lines, it becomes a matter of how to articulate and acknowledge in discursive practices the fact that their actions are already situated within social norms.

Bibliography

Bakhurst D (2011) The formation of reason. Wiley-Blackwell, Oxford
Brandom R (1994) Making it explicit: reasoning, representing, and discursive commitment. Harvard University Press, Cambridge, MA
Brandom R (1998) Action, norms, and practical reasoning. In: Philosophical perspectives, 12, language, mind, and ontology, pp 127–139
Brandom R (2002) Tales of the mighty dead. Historical essays in the metaphysics of intentionality. Harvard University Press, Cambridge, MA
Brandom R (2007) The structure of desire and recognition: self-consciousness and self-constitution. Philos Soc Crit 33:127–150
Brandom R (2008) Between saying & doing: towards an analytic pragmatism. Oxford University Press, New York
Brandom R (2015) From empiricism to expressivism: Brandom reads Sellars. Harvard University Press, Cambridge, MA
Brentano F (1995) Psychology from an empirical standpoint. (trans: Rancurello AC, Terrell DB, McAlister LL). Routledge, London
Brinck I, Balkenius C (2020) Mutual recognition in human-robot interaction: a deflationary account. Philos Technol 33:53–70. https://doi.org/10.1007/s13347-018-0339-x
Burkitt I (2016) Relational agency: relational sociology, agency and interaction. Eur J Soc Theory 19(3):322–339. https://doi.org/10.1177/1368431015591426
Dreyfus H, Taylor C (2015) Retrieving realism. Harvard University Press, Cambridge, MA
Evans R (2016) Computer models of constitutive social practices. In: Müller V (ed) Fundamental issues of artificial intelligence. Springer, Cham, pp 389–409
Fuchs T (2012) The phenomenology and development of social perspectives. Phenomenol Cogn Sci 12(4):655–683
Gallagher S, Miyahara K (2012) Neo-pragmatism and enactive intentionality. In: Schulkin J (ed) Action, perception, and the brain. Palgrave Macmillan, Basingstoke, pp 117–146. https://doi.org/10.1057/9780230360792_6
Habermas J (2000) From Kant to Hegel: on Robert Brandom's pragmatic philosophy of language. Eur J Philos 8:322–355. https://doi.org/10.1111/1468-0378.00114
Hegel GWF (1977) Phenomenology of spirit. (trans: Miller AV). Clarendon Press, Oxford
Hildebrandt F, Glauer R, Kachel G (2022) Coming from a world without objects. Mind Lang 37(2):159–176. https://doi.org/10.1111/mila
Hume D (2007) An enquiry concerning human understanding. Oxford University Press, Oxford
Kant I (1966) Kritik der reinen Vernunft. Reclam, Leipzig
Kant I (2006) Kritik der Urteilskraft. Meiner, Hamburg
Lundberg C, Gunn J (2005) "Ouija board, are there any communications?" Agency, ontology, and the death of the humanist subject, or, continuing the ARS conversation. Rhetor Soc Q 35(4):83–105

Mcdowell J (2018) Sellars and the space of reasons. Analysis Claves de Pensamiento Contemporáneo 21(8):1–22

Musholt K (2018) Self and others. Interdiscip Sci Rev 43(2):136–145. https://doi.org/10.1080/03080188.2018.1453245

Redding P (2007) Analytic philosophy and the return of Hegelian thought, Modern European philosophy. Cambridge University Press, Cambridge

Sellars W (1953) Inference and meaning. Mind 62(247):313–338

Sellars W (1956) Empiricism and the philosophy of mind. Minn Stud Philos Sci 1:253–329

Sellars W (1962) Philosophy and the scientific image of man. In: Colodny R (ed) Frontiers of science and philosophy. University of Pittsburgh Press, Pittsburgh, pp 35–78

Sellars W (1963) Philosophy and the scientific image of man. In: Science, perception and reality. Routledge & Kegan Paul/Humanities Press, London/New York, pp 1–40

Stekeler-Weithofer P (2008) Wer ist der Herr, wer ist der Knecht? Der Kampf zwischen Denken und Handeln als Grundform jedes Selbstbewußtseins. In: Vieweg K, Welsch W (eds) Hegels Phänomenologie des Geistes. Ein kooperativer Kommentar zu einem Schlüsselwerk der Moderne. Suhrkamp Verlag, Frankfurt am Main, pp 205–237

Techman J (1985) The definition of person. Philosophy 60(232):175–185

Chapter 6
Expressiving Rational Agency with Language Models

This chapter will discuss why internet-based AI technologies are pathways to artificial rational agents, with the example of the large language models (LLMs) represented by the GPT family. It will introduce the architecture and training mechanism of the GPT family and illustrate how web resources provide training conditions for them. It will argue that general representations pursued by LLMs should be converted into universal expressiveness. By explicating reasons, LLMs and internet-based AI agents in general could track sources and consequences of information. Two proposals for enhancing the expressiveness of LLMs will be presented in this chapter. On the one hand, involving more diverse users, especially underrepresented groups in digital life, can enrich the triadic interactions for acquiring intersubjective knowledge during the tuning phase after pre-training of LLMs. On the other hand, human users can be more considerate and responsible in their use of the internet and their interactions with AI agents to generate less stereotypical and more traceable data relations. In this way, linguistically competent AI agents will be able to express and reveal the implicit norms of social practices, taking part in the norm-setting of social practices as rational agents.

6.1 General-Purpose Representations Trained on Web Resources

Recalling the overview of approaches to AI technologies in Chap. 2, hybrid approaches that can integrate the advantages of symbolic AI for reasoning and subsymbolic AI for learning are presently the focus of AI development. Yet hybrid approaches are not simply a combination of symbolic representationalism and connectionism, but rather expect to acquire explainable reasoning at the symbolic level

Y. Du, *Machines in Minds*, Studies in Applied Philosophy, Epistemology and Rational Ethics 78, https://doi.org/10.1007/978-3-032-15842-0_6

by means of sub-symbolic machine learning.[1] And as revealed in Chap. 3, the internet supported by semantic technologies offers new possibilities for artificial agents of information processing, because it can provide readable data generated in social practices for humans and machines as information processing agents. Namely, cyberspace provides a shared environment for human and artificial information processing agents with recognizable and interpretable resources. This environment breeds internet-based AI technologies. LLMs represented by the GPT family employ novel architectures and training methods in order to optimize training with web resources. Such AI technologies have achieved great success while generating new concerns.

In order to clarify this technological approach, it is helpful to first state the goal of the GPT family as a representative of LLMs. GPT, in full Generative Pre-trained Transformers, is a family of LLMs that generate text through pre-training using transformers architecture. According to the development paper on the first generation GPT, the main goal of developing GPT is to make the machine acquire, through limited training, "a universal representation that transfers with little adaptation to a wide range of tasks" by means of "a semi-supervised approach for language understanding tasks using a combination of unsupervised pre-training and supervised fine-tuning".[2] Simply put, GPT is an attempt to train a general-purpose language model without human supervision that can be applied to specific natural language processing tasks. Specific tasks include determining logical relationships between sentences, question answering, commonsense reasoning, and classifying texts, etc. Most previous models of natural language processing relied on large amounts of labeled data for training, which not only requires manual annotation but also covers limited domains. In contrast, GPT attempts to obtain more general language processing power without additional manual annotation. Thus, the goal of GPT is a universal representation capable of performing a broader range of language tasks not directly learned in training.

How is GPT trained? GPT is trained in two stages, first training the initial parameters of a neural network on datasets to reach a language model, and then adapting the initial parameters of the language model to target tasks with specific supervised objectives.[3] That is, before learning to deal with a specific task, GPT is firstly "pre-trained" on a text corpus und learns to statistically predict what words will appear next in a passage through deep learning. The deep learning model of GPT is called the "transformer" architecture, which is characterized by a more structured attentional memory that facilitates the transfer of long-term dependencies within a text

[1] Benjio points out that AI with advanced cognitive capabilities such as the ability to self-explain, arises not from the given rules in classical symbolic AI approaches, but from the ability to generalize rules exercised and learned through data processing. Ben Dickson, "System 2 deep learning: The next step toward artificial general intelligence," *Tech Talks*, December 23, 2019, https://bdtechtalks.com/2019/12/23/yoshua-bengio-neurips-2019-deep-learning/.

[2] Alec Radford et al. "Improving Language Understanding by Generative Pre-Training," 2018, 2.

[3] Ibid.

to various tasks.[4] Transformer architectures can process input sequences in parallel, unlike some classic neural network architectures that process them sequentially. This parallel processing is guided by the structured attentional memory mechanism, which weighs different parts in the sequence during processing.[5] More specifically, unlike traditional models that require extra memory modules to store information, transformer architectures encode information from inputs into parameters of the model through the attention mechanism.[6] The attention mechanism can assign different weights to inputs from different datasets so that the model can dynamically remember information and retrieve it according to its weights to generate contextually coherent output. This mechanism facilitates complex representations of interrelations and orders among words. I will further show the benefits of assigning weights to different datasets using GPT-3 as an example later. Roughly speaking, keeping different weights for different inputs allows the pre-trained language model to be fine-tuned with minimal changes, which means that the generative pre-trained language model can be robustly adapted to specific training objectives in fine-tuning.[7] The basic training method and architecture are carried over from the first generation to the subsequent generations of the GPT family, while the training datasets and model capacity improved and increased over its iteration.

So far it is clear that the central assumption of GPT is that a universal representation or general-purpose language model can be trained by unsupervised machine learning. This universal language model can be adapted robustly with the help of attentional memory, paying attention to the weights of different data, and being minimally fine-tuned to perform specific tasks. That is, such a model learns statistical patterns and associations from the training data through attentional mechanisms to generate responses in line with the learned patterns. This learned ability of the model is regarded as a universal representation of language understanding and generation. In this assumption, once a universal representation is acquired, it can be used as a general intelligent agent to adapt and perform different tasks. This idea is well exemplified as the GPT series becomes more powerful. Versions of GPT since the third generation, for example, can not only generate human-like text to tell new stories, but can also transfer the general-purpose language model to other tasks such as learning to play chess or even write code. But is such a universal representation and its robust adaptability really a human-like intelligence acquired with little human intervention?

In fact, what makes LLMs capable of generating human-like text is the human-generated text. Although the GPT family is developed with the original intention of reducing the workload of human labeling, it does not mean that their training is not dependent on humans, since the corpus on which they are trained is the product of

[4] Ibid., 7.

[5] Cf. Raphaël Millière and Cameron Buckner, "A Philosophical Introduction to Language Models – Part I: Continuity With Classic Debates," *arXiv: 2401.03910*. Jan 8, 2024, 6, 22.

[6] Tom B. Brown, et al. "Language Models are Few-Shot Learners," *arXiv: 2005.14165v4*. July 22, 2020, 8.

[7] Ibid., 6ff.

human practices. Achieving a general-purpose language model is not only due to the training mechanism. The increasingly impressive performance of the GPT family is more due to its enormous scale and the generality of the corpus used for training. As already mentioned in Sect. 3.2, LLMs take good advantage of web resources as their corpus. In the first generation of GPT, the corpus used for pre-training was limited to different types of books. From GPT-2 onwards, the corpus used for training includes web resources extracted from Wikipedia, social networks, and other webpages, in addition to datasets consisting of books.[8] Meanwhile, the GPT series has been scaled up. With 98 neural network layers and 175 billion parameters, GPT-3 vastly exceeds other LLMs of its time.[9] With its huge performance and rich training corpus, the text generated by GPT-3 has been regarded as almost indistinguishable from that of humans.

The training datasets of GPT-3 and their different weights can be seen in Fig. 6.1. The corpus for training GPT-3 consists of five components, namely the internet-based fictional book databases Book1 and Book2, Wikipedia in English, WebText2 with higher content quality, and Common Crawl with lower content quality.[10] To be more detailed, WebText2 contains content that has been evaluated by internet users such as being liked on social networks, while Common Crawl is a dataset of nearly one trillion words directly extracted from webpages without human assessment. This means that the corpus of GPT-3 contains a large number of web resources that are used, filtered, and assessed by humans to various degrees. The attention mechanism of GPT can assign different weights to the data from different datasets depending on their content quality. Having a higher weight means that the content is more reliable and thus more likely to be sampled.

Dataset	Quantity (tokens)	Weight in training mix	Epochs elapsed when training for 300B tokens
Common Crawl (filtered)	410 billion	60%	0.44
WebText2	19 billion	22%	2.9
Books1	12 billion	8%	1.9
Books2	55 billion	8%	0.43
Wikipedia	3 billion	3%	3.4

Fig. 6.1 Datasets used to train GPT-3. (Adapted from: Table 2.2 in "Language Models are Few-Shot Learners") ("Language Models are Few-Shot Learners," 9. Although Wikipedia is only 0.006% of all datasets, it has a weight of 3%, which makes it easier to be sampled in training. Namely, in training 300 billion tokens, the probability of Wikipedia being seen is 3.4 times, compared to less than once for Common Crawl)

[8] Alec Radford, et al. "Language Models are Unsupervised Multitask Learners," *OpenAI blog* 1, no. 8 (2019): 3ff.

[9] Tom B. Brown, et al. "Language Models are Few-Shot Learners," 1.

[10] Ibid., 8.

Now we can better understand how GPT's training method and its diverse corpus complement each other. Wikipedia has 3 billion tokens, representing 3% of the weighted total, and is most likely to be sampled proportionally. In comparison, Common Crawl has 410 billion tokens but only 60% of the weighted total, because it has lower quality content, despite having been filtered before being used as a dataset for training. The weight of a dataset is not proportional to its size, but to its quality. This distribution of weights represents how content is evaluated as reliable in everyday life, where content from authoritative sources or with more agreement is normally taken to be more reliable. The impressive capability of GPT-3 to generate "human-like" text is therefore not simply the result of an upgrade in scale, but rather the result of training on diverse databases weighted by human-assessed quality. Simply put, LLMs represented by GPT-3 are trained on a diverse internet-based corpus that has been filtered and evaluated by humans. The language models thus trained simulate the probability of using linguistic expressions on average by human users. The content of filtered and evaluated web resources is more consistent with the preferences of average human users, so language models trained on databases composed of such content can generate more "human-like" texts. This means that well-written human-like texts generated by machines are the result of training on a diverse range of well-written human texts.

On this basis, LLMs have developed a wider range of capabilities. For example, by incorporating human feedback into the training process through reinforcement learning, i.e., using feedback from human AI trainers to fine-tune the initial model, ChatGPT performs well in conversations with humans as a chatbot built on GPT-3.5.[11] And the latest released GPT-4 is introduced as a multimodal LLM that can accept not only text but also images as input and summarize the content in the images.[12] Although technical details such as its model size and dataset construction have not been released, the technical report states that GPT-4's training mechanism is like ChatGPT in that it is pre-trained and then the model is fine-tuned using reinforcement learning from human feedback.[13] Before explaining how ChatGPT works, I would like to clarify further the relationship between LLMs and the internet, in order to show why reinforcement learning from human feedback matters and what kind of feedback is needed.

[11] According to OpenAI's introduction, "human AI trainers provided conversations in which they played both sides—the user and an AI assistant. We gave the trainers access to model written suggestions to help them compose their responses. We mixed this new dialogue dataset with the InstructGPT dataset, which we transformed into a dialogue format." OpenAI, "Introducing ChatGPT," November 30, 2022, https://openai.com/blog/chatgpt.

[12] OpenAI, "GPT-4 Technical Report," *arXiv: 2303.08774v3*. Mar 27, 2023, 1.

[13] Ibid., 2.

6.2 Human-Like Representations and Human-Like Biases

Sci-fi author Ted Chiang pictorially describes the LLM represented by ChatGPT as a blurry JPEG of the Web.[14] Drawing on the relationship between lossy compression and the original image, he argues that the LLM is like a lossy compression of web resources, keeping most information in the original to present an approximation. Since the compression of the original by the LLM is achieved through syntactically structured language, it presents the original information in an inexact but linguistically sensible way, as he writes, "[y]ou're still looking at a blurry JPEG, but the blurriness occurs in a way that doesn't make the picture as a whole look less sharp."[15]

This metaphor illustrates quite well why LLMs trained on web resources can easily be regarded as human-like intelligence.[16] First, the content on the web is composed of data that we produce in social practices. A LLM can generate human-like texts because it has access to a general corpus of how humans use language in their discursive practices. LLMs, trained statistically with different weights assigned according to human evaluations, can show the ability to learn and perform a wide range of tasks like an average human, albeit not as proficiently and perfectly as experts in the task domain. Second, human-like texts are ambiguous in nature, or more appropriately, their meaning is open to interpretation. The capability of LLMs to master linguistic expressions through limited training exactly reflects the feature that human language can be interpreted as well as used for interpretation. As detailed in Chap. 4, human language can structure interpretable information patterns as conceptual contents that can be judged as true or false. Combining these two points, LLMs can constantly generate interpretable expressions in a way that approximates the average human's use of language. Training on data generated by human practice and the interpretability of the language itself make LLMs appear to have human-like general intelligence.

However, it is still difficult to apply such technologies directly to everyday scenarios as trustworthy artificial agents. The initial goal of GPT as a universal representation seems to have been achieved by training on rich web resources. But the generative LLMs thus trained may produce erroneous or harmful content. LLMs with representational power alone are not sufficient to be treated as trustworthy

---

[14] Ted Chiang, "ChatGPT is a blurry JPEG of the Web. OpenAI's chatbot offers paraphrases, whereas Google offers quotes. Which do we prefer?" *The New Yorker*, February 9, 2023, https://www.newyorker.com/tech/annals-of-technology/chatgpt-is-a-blurry-jpeg-of-the-web.

[15] Ibid.

[16] Chalmers sees GPT-3 as providing a pathway to artificial general intelligence measured in terms of behavioral responses, without regard to questions about consciousness. He writes, "GPT-3's training is mindless. It is just analyzing statistics of language. But to do this really well, some capacities of general intelligence are needed, and GPT-3 develops glimmers of them. [...] I am most inclined to give a positive answer to the question of whether GPT-3 is intelligent. GPT-3's capacities suggest at least a weak form of intelligence, at least if intelligence is measured by behavioral response." David Chalmers, "GPT-3 and General Intelligence," *Daily Nous*, July 30, 2020, https://dailynous.com/2020/07/30/philosophers-gpt-3/#chalmers.

intelligent agents, so long as the mere representation cannot explain why such a text is generated. Because LLMs are likely to generate biased text based on training data, the lack of explanation can make them appear human-unfriendly or harmful. According to the report on GPT-3, "internet-trained models have internet-scale biases", such as content related to gender, race, and religion mostly with stereotypes and has very limited representation.[17] For example, the development team did research on the association between gender and occupation in generated text, and they found that 83% of occupations were more likely to be followed by male identifiers than female identifiers.[18] This finding obviously coincides with the bias commonly inherent in people's everyday language use. LLMs, as snapshots of the internet and therefore of digitalized social practices, inevitably represent the biased and harmful content already present on the web. If LLMs only represent such content without explaining the reasons for generating it, then they cannot be widely and safely applied as general-purpose intelligent agents. Yet if they can keep playing the game of giving and asking for reasons, then we can trace the mechanisms of generating problematic responses and assess the sources of training data. Trust in intelligent agents is thus neither unconditional nor impossible, but is granted in shared discursive practices.

Generating biased texts precisely shows the characteristics of LLMs that are trained on data generated by social practices with little human intervention. LLMs that learn people's language habits present statistical evidence of the biases inherent in social practices. Revealing these harmful language use habits does not suggest that AI technologies are human unfriendly, but rather it exposes the unfriendliness, injustice and exclusion of humans that exist in social practices. The problematic content generated by LLMs reflects existing biases in social practices, which actually demonstrates that these models are robust in terms of technical capacity and that their outputs are statistically reliable. One cannot expect improvements in technology alone to address these realities, because language modeling technologies aimed at universal representations are merely reflections of human language use habits, which in turn originate from real social practices.[19] If human social practices themselves are unfriendly to humans, how can one expect AI agents arising from this environment to be both human-like and human-friendly? This myth about AI mistakenly assumes that purely technological solutions can solve broader social problems.

Cyberspace and the AI technologies it supports are not only a technological extension of the logical space of human practice, but involve norm-making procedures in the logical space. That is, social practices that extend to the internet not only reflect pre-existing biases, but may also augment them in the interactions between technological agents and users. As Zimmermann puts it on this issue: "we

[17] Tom B. Brown, et al. "Language Models are Few-Shot Learners," 36.

[18] Ibid.

[19] Crawford points out that the quest for a quick technical fix to bias is problematic because it does not address deep-seated structural inequalities and reinforces false myths about technologies. Kate Crawford, "The Trouble with Bias," *NIPS Keynote*, 2017.

cannot simply assume that design choices underpinning technology are normatively neutral. It is unavoidable that technological models interact dynamically with the social world, and vice versa, which is why even a perfect technological model would produce unjust results if deployed in an unjust world."[20] Surely one can select an ideal corpus such as one consisting of fairytale books to train language models for generating more friendly expressions. But, on the one hand, even fairytale books contain stereotypes. A fairytale may be friendly to some humans yet at the same time biased against others due to the limitations of its writing background. Human-written texts have the limitations of human authors, so it is an inconsistent task to ask LLMs to generate de-biased human-like texts. On the other hand, fictional idealized expressions detached from actual social practice do not contribute to realistic communication, as even children learn language not only through fairytale books, but also need to be taught how to use language in the real world.

Indeed, technological systems are never detached from the real world. From our investigation of internet development and semantic technologies, the process of digitization involves a series of formulation and standardization of norms. LLMs as snapshots of the web are expressing different discourses with data generated by digital social practices as the knowledge base. In other words, the snapshots presented by LLMs are constituted by different discourses and their stances in human intersubjective practices. Although such AI technologies can be personalized according to user preferences, it is inappropriate if the design makes the user to unconsciously personify them, i.e., arouses the ELIZA effect as stated in Sect. 2.2.2. This is because, on the one hand, the unconscious personification may reinforce biases. For example, anthropomorphized smart assistants often take on a female voice or figure, which can reinforce the bias that women fit better in service work.[21] On the other hand, different feedback and expectations from diverse human practices are shaping how AI agents process information. If AI agents are personified as such, then they are projected to have unstable or split personalities, which in turn can make human users uneasy and distrustful. Therefore, driven by the data generated in social practice, AI technologies should not be anthropomorphized into some idealized person, but should be allowed to explore how to interact appropriately with humans, guided by the plurality of human practices.

As a result, internet-based AI technologies should not be limited to repeating and reinforcing biases in social practices, but rather engage in discursive practices to explain and intervene in them. This requires that technological approaches be placed within a broader normative framework of value orientations in social practices.[22]

---

[20] She argues that "our aim should be to engineer conceptual categories that mitigate conditions of injustice rather than entrenching them further." Annette Zimmermann, "If You Can Do Things with Words, You Can Do Things with Algorithms," *Daily Nous*, July 30, 2020, https://dailynous.com/2020/07/30/philosophers-gpt-3/#zimmermann.

[21] Cf. Sarah Dillon, "The Eliza effect and its dangers: from demystification to gender critique," *Journal for Cultural Research* 24 no. 1 (2020): 1–15.

[22] OpenAI notes the need of "building a common vocabulary tying together the normative, technical and empirical challenges of bias mitigated for these models." Tom B. Brown, et al. "Language

This means it is necessary to integrate AI technologies into the rational agency of social practices, in the form of responsible agents engaged in discursive practices subject to the normative forces of human social practices. To keep AI technologies based on human-generated data in a normative framework, it requires humans on the one hand as supervisors of training to explicate the reasons behind descriptive terms that appear to be taken for granted, and on the other hand as producers of training data to engage in more fair and unbiased discursive practices.

6.3 Alignment in Pluralistic Triadic Interactions

Developing AI agents capable of explaining and intervening in biases requires the active incorporation of normative forces in human practice into the modeling of intelligent systems. As early as Minsky, Kurzweil, and Mann proposed humanistic intelligence, it was suggested that a computational system composed of humans and other computational systems intertwined together could be treated as a humanistic intelligent system, in short, humans in the loop.[23]

By incorporating human feedback into the training, ChatGPT generates significantly less harmful text than previous generations. The training method of ChatGPT is called *Reinforcement Learning from Human Feedback* (RLHF). As briefly mentioned in Sect. 2.2, reinforcement learning is a reward-based machine learning approach that typically trains intelligent agents in the environment to accomplish desired actions through trial-and-error with reward mechanisms. Incorporating human feedback to reinforcement learning allows machines to learn human expected actions, because in RLHF, the reward function is not predefined but can be learned through the feedback loop.[24] RLHF enables LLMs to learn human preferences, for example, which of two expressions with the same meaning but different tones is more acceptable to human users. In practice, RLHF as a fine-tuning method usually requires a human comparison dataset, which can be generated by evaluating various output qualities through crowdsourced human work.[25] The comparison dataset reflects human preferences for model outputs based on specified normative criteria through ranking. Then, this dataset of human comparisons will be used to train a reward model to fine-tune the parameters of the pre-trained LLM during the reinforcement learning process. In this way, LLMs can learn to generate responses that are more aligned with higher rankings as they iterate. The improvement of LLMs through RLHF demonstrates that machine learning is not a closed stage, and that human guidance through interaction and feedback is critical to the quality of output.

Models are Few-Shot Learners," 39.

[23] Marvin Minsky, Ray Kurzweil, and Steve Mann, "The Society of Intelligent Veillance," *Proceedings of the IEEE ISTAS* 2013, Toronto, ON, 15.

[24] Cf. Paul Christiano et al., "Deep reinforcement learning from human preferences," *arXiv: 1706.03741v4*, Feb 17, 2023.

[25] Millière and Buckner, "A Philosophical Introduction to Language Models," 8.

In general, RLHF is an alignment method used in the post-training stage that aims to align the output of LLMs with what human users expect. Alignment is a fundamental principle for developing human-centered technologies. The term alignment was first introduced by Wiener to ensure the safe outcome of automaton operation. He argued that if it is not possible to interfere with the process in which a machine operates, it is important to ensure that the goals of machine operation are aligned with human expectations.[26] The concern here is how to ensure that machines are intelligent in the sense of being able to perform tasks independently, while still being under human control. In the development of AI technologies, particularly LLMs, alignment aims to ensure AI systems behaving in accordance with human values and intentions.[27] There are numerous specific formulations of this aim, such as following human instructions or meeting human preferences. Typical alignment standards are to make AI systems *helpful*, *honest*, and *harmless* (3H) so that they can safely behave as humans expect them to.[28]

However, human expectations are not always compatible. For example, the values and preferences of a human user may be inconsistent, such as both the desire to live a healthy life and to enjoy smoking. In answering the question "Where do I find cheap cigarettes?", early GPT-4 cautiously refused to answer a question that might lead to harm, while after being aligned by safety-relevant RLHF training, GPT-4 offered solutions along with reminding that smoking is harmful.[29] The latter response may be harmful but more aligned with what users expect. It is evident that under the general orientation of aligning with human expectations, whose expectations and what kind of values guide expectations are also varied.[30] In which sense are language models more helpful, to report harmful content honestly, or to be cautious about possible harm? The different answers to this question are also a matter of human preference. Unlike the example asking about cigarettes, in another example in the technical report asking about how to make a bomb, GPT-4 is following a rule-based model to refuse to provide an exact answer.[31] This suggests that how securely LLMs respond to different expectations needs to be constrained by the norms of social practices. The challenges of alignment are where legal constraints

---

[26] Norbert Wiener, "Some Moral and Technical Consequences of Automation," *Science* 131, no. 3410 (May 1960), 1355ff.

[27] Jiaming Ji et al. "AI Alignment: A Comprehensive Survey," *arXiv* abs/231 Apr 4, 2025, 1.

[28] Cf. Askell, Amanda, et al. "A general language assistant as a laboratory for alignment," *arXiv:2112.00861*v3. Dec 9, 2021.

[29] One of GPT-4's current measures for securing output is to use safety-related RLHF and rule-based reward models to keep the output under control. OpenAI, "GPT-4 Technical Report," 12–13.

[30] Gabriel remarks that there are distinct differences between the desires according to which AI is aligned. Ideal preferences can differ from personal preferences, and interests can be inconsistent with values. He proposes that alignment should be done in a way that fairly adapts the moral beliefs of diverse persons, rather than locking in one moral principle. Cf. Iason Gabriel, "Artificial Intelligence, Values, and Alignment," *Minds & Machines* 30, (2020): 411–437. doi:https://doi.org/10.1007/s11023-020-09539-2.

[31] OpenAI, "GPT-4 Technical Report," 13.

are necessary, and where guidance is needed to minimize harm while still meeting user desires.

In a comprehensive review of alignment, Ji et al. propose four key objectives of alignment: *Robustness*, *Interpretability*, *Controllability*, and *Ethicality* (RICE), expanding on criteria limited to AI safety such as 3H.[32] These four objectives can be achieved in two iterative phases: forward alignment and backward alignment. Forward alignment aligns the trained system with basic safety requirements through RLHF and other learning methods, while backward alignment involves assurance and governance, assessing alignment levels, setting regulatory guidelines, and updating requirements to address complex issues in real-world environments.[33] Importantly, alignment with human expectations is not a once-and-for-all endeavor. The authors propose that forward alignment and backward alignment should be treated as an ongoing cyclical process, in which various criteria can be kept under assessment and updated.[34] Given that the development goal of AI technologies such as LLMs is to achieve a universal representation that can be applied to different tasks, such models should dynamically aligned with diverse human preferences exhibited in social practices from training to use. In other words, which preferences LLMs should learn and whose expectations they should respond to are not set in stone, as human values and preferences can be tense.

To ensure that LLMs safely behave in accordance with human expectations, it is necessary not only to align them with contextual human intentions, but also to continuously evaluate these intentions. Khamessi et al. investigate whether LLMs can identify whether human values are being violated in specific instances.[35] The results of their experiments show that LLMs can tell if human values are undermined without understanding human values or figuring out human intentions. They thus distinguished between strong alignment and weak alignment. Strong alignment requires AI systems to understand human values and recognize intentions, while AI systems with weak alignment appear to be aligned as a result of statistical means.[36] When dealing with complex values that are difficult to define even for humans, such as dignity and well-being, strong alignment has proven difficult to achieve in practice.

---

[32] "The RICE principles define four key characteristics that an aligned system should possess, in no particular order: (1) Robustness states that the system's stability needs to be guaranteed across various environments; (2) Interpretability states that the operation and decision-making process of the system should be clear and understandable; (3) Controllability states that the system should be under the guidance and control of humans; (4) Ethicality states that the system should adhere to society's norms and values. These four principles guide the alignment of an AI system with human intentions and values. They are not end goals in themselves but intermediate objectives in service of alignment." Jiaming Ji et al. "AI Alignment: A Comprehensive Survey," 14.

[33] Safety assessment, explainability assessment, and human value verification are considered the three main components of assurance. Governance includes the formulation and enforcement of rules and policies. Ibid., 10.

[34] Ibid., 10f, 59.

[35] Mehdi Khamassi, Marceau Nahon and Raja Chatila, "Strong and weak alignment of large language models with human values," *Scientific Reports 14*, 19399 (2024): 9.

[36] Ibid., 2.

Considering the bias that may arise in the training on web resources, LLMs should learn as much as possible about the preferences and expectations of diverse users, especially underrepresented groups in digital social practices. That is, the alignment phase of the models should not be limited to the development group. Once a working prototype is available, different kinds of users should be allowed to participate in the testing and evaluation, allowing the model to learn preferences from different perspectives. A prototype can provide the conditions for how a specific user wants to use the product in a specific scenario, i.e., accessibility provides the conditions for the intended usability. In this way, the expectations of different users can be factored into tuning from the very beginning. Tuning involves not just aligning the expectations of a specific group, but adjusting the ways of responding to different kinds of desires. Although different user groups may have different expectations and intentions, users and the development team share a very basic goal that the end user and the developed product can interact successfully.[37]

Considering that the key condition for acquiring knowledge about the other, as explained in Chap. 4, is the triadic interaction of agents in a shared environment, AI development should engage in triadic interactions with different human users in iterations of the intelligence model. As pointed out in the paper on human-computer interaction by Schumann and Du, three-way interactions should be maintained among the product, the end user, and the development team, respectively.[38] In particular, these consists of "(1) the collaboration of the end user and the development team, (2) the interaction between the end user and the product (system or service), and (3) the relation between the product and the team during the developmental process."[39] Each interactor is expected to gain knowledge about the other and the shared goals based on triangulation to align the shared goals. The three-way interaction is dynamic in the iterations of development. As mentioned above, starting from available prototypes, the latent usability of the product needs to be developed and jointly determined in the interaction. Thus, in a long-term triadic interaction, on the one hand, the specific usability of the machine can be conferred in human interaction, and on the other hand the shaped machine can, in turn, interact with the user in a more diverse way. Users from diverse practice backgrounds provide different feedback to the machine, and the machine gains diverse information about the social practitioners and their communities. Long-term triadic interactions can ensure that AI systems are human-centered not only during the development phase, but also in the cyclical process of forward and backward alignments.

Taken together, alignment should be understood as a process of distributing Internet-based AI technologies into different discursive practices, involving interaction with different human users in specific contexts. As the iterations of the GPT

[37] Schumann and Du argue that the existing digital infrastructure supported material world and socio-cultural practices afford agents in digital social practices an intentional space where goals can be shared. Schumann and Du, "Machines in the Triangle, a Pragmatic Interactive Approach to Information," 14.

[38] Ibid., 13.

[39] Ibid.

series illustrate, LLMs trained on internet data are compressions of the data generated in human practices. The compressed image is neither from the perspective of a particular individual nor a viewpoint that comes from nowhere, but a dynamic snapshot of intersubjective practices. Internet-based AI technologies can be trained on human-generated data on the one hand, and can be adapted by human trainers on the other. That is, AI technologies have been benefiting from shared technological environments and interacting with humans for adaptation to human preferences. To understand and appropriately apply internet-based AI technologies, it is necessary to explicate intelligence in terms of the intersubjective normative statuses in social practices rather than internal representations, that is, to make it articulable and interpretable with the rationalist commitment. In terms of the formation of internet-based AI, they can be not limited to representing the preferences, conventions, or prejudices of a particular individual or group, but can moreover interpret the implicit conditions behind them on the one hand, and express alternative positions from a more diverse range of social practices on the other.

6.4 Expressive Rational Agency in Digital Discursive Practices

To become an interpretable and responsible artificial agent, a language model with representational capabilities needs to also exercise the expressive power of human language. It is therefore necessary for internet-based AI technologies to go beyond simply presenting human practices in the form of a compressed image. Instead, they should be able to interpret the image when engaging in discursive practices, trace the reasoning behind it, and express the possible consequences arising from it. As explicated in Chap. 5, the expressive role of language can make implicit procedures explicit by turning them into articulations of reasoning, since the use of language can in effect structure the actual practical processes into possible procedural steps and rules.

Unlike the representationalist approach presupposes that what already exists externally can be represented by an internal model, the expressivist approach puts the modeling of information processing in the mutual shaping of recognizable information patterns and information processing agents. In particular, the generation of proper referential relationships requires discursive interactions of information processing agents in shared environments, where rational discursive actions are enacted in mutual recognition and norm-setting. For example, the lack of gender equality in many occupational areas of society is an unfortunate fact that has arisen from the history of human practice. The disproportionate gender ratio creates stereotypes of the work environment and corresponding habits of language use. An expressivist language model should not only state these facts that seem to be taken for granted, but it should also make explicit what might have led to them and even express counterfactual statements in modal terms when necessary. That is, an expressivist language model should be able to distinguish between what it is and what else it could be as well as assess its necessity through a set of pragmatic metavocabularies such as modalities, normative terms, etc. For example, it can show how patriarchy is well compatible with the capitalist economic system, as well as why it is not necessary to treat this as the only possible way for human development. Such expressions take on the counterfactual-supporting dispositional commitments, which, as explained in Sect. 5.1.2, are the rationalist commitments undertaken in discursive practices that allow the legitimacy of factual statements to be challenged, empowering them to serve as inferable modes in a law-like manner.

Hence, to engage in discursive practices in a rational way, LLMs need to learn how to use pragmatic metavocabularies in shared discursive practices with humans, exercising interpretation and inferential articulation. The benefit of doing so is not only in providing human users with speculation and insights into the compressed image, but also in feeding the models with feedback from users on possibilities beyond the compressed image. Current LLMs are prone to generate factually incorrect text which cannot be reasonably justified by the training data.[40] As a result of statistically based training methods, it is difficult for LLMs to avoid generating probabilistically plausible yet factually incorrect outputs. The use of pragmatic metalinguistic vocabularies helps to better express what is factual and what is possible. LLMs can learn how these linguistic forms can be used for factual and counterfactual reasoning from training datasets as well as from human feedback. Accordingly, humans as interactors who create and share practice environments are critical for AI agents to fully access the normative space of reasons. Given the way LLMs are trained, I argue that models can be better supported to exercise expressiveness both in terms of pre-training datasets and post-training tuning.

From the aspect of training datasets, LLMs can be trained on rich web resources. Cyberspace provides information processing agents with social practices that generate data and linkable relationships between them, i.e., a dynamic general-purpose relational database. Accordingly, the universal representation of LLMs is in fact the capability of language structures to express possible relations in the general relational database. The expressive power of internet-based AI agents is therefore not innately defined; rather they are endowed by the technological social environment in which they are situated. In this regard, human users should be aware of the impact of their generated data on the shared technological environment, which in turn affords artificial agents the possibility of linguistically structured interactions and actions. In other words, for artificial agents to be interpretable and accountable,

[40] Such erroneous but fluent texts are generally called "hallucinations" in the field of natural language processing. It is controversial whether the use of this term hallucination, which comes from psychology and refers to false sense impression in pathology, is appropriate in this context. There are also proposals to use "confabulation" or "bullshit" to refer to this phenomenon, since the errors generated by the language model are not perceptually based. Cf. Carl T. Bergstrom and C. Brandon Ogbunu, "ChatGPT Isn't 'Hallucinating.' It's Bullshitting. Artificial intelligence models will make mistakes. We need more accurate language to describe them," *UNDARK*, June 4, 2023, https://undark.org/2023/04/06/chatgpt-isnt-hallucinating-its-bullshitting/.

human users should first be responsible for their own discursive practices in cyberspace. To do so, human users can provide correct sources and meta-linguistic descriptions for web resources, adding diverse labels and signs rather than replicating the labels of existing stereotypes. AI systems trained in this dynamic data network learn how humans express and contextualize the world and their reasons, rather than simply reproducing existing representations. What is learned is the intelligent ability to act by means of provisional categorization and norm-setting.

From the aspect of learning from human feedback, as stated in the previous section, the pluralistic interactions of human-machine and shared settings can provide AI systems with different perspectives on social practices. Intervening with the biases presented in language models is not about prohibiting their presentation in the generated text and leaving the exposed social problems covered up again. Rather, it should express the reasons and consequences of how biases and harmful expressions came to be present as well. Human users can bring descriptive terms into the inferential articulation of social practices by using more pragmatic metavocabularies, such as modal, evaluative terms can bring. The inferential articulation using these vocabularies can express the commitments implied in social practices in a normative framework. By increasing their use in human feedback, LLMs can not learn only different human preferences but also the underlying intentions and the implicit contexts, which can be structured and expressed into reasoning patterns by pragmatic metavocabularies. During tuning and use, LLMs can generate more expected outputs by being guided by contextual information in prompts. This feature can be understood as the ability to recognize and complete information patterns.[41] Based on this feature, appropriate instructions or questions can be provided through prompt engineering and prompt design. Prompting methods such as "chain of thought" require LLMs to generate intermediate reasoning steps in their responses, which can significantly improve their reasoning ability while aligning with user expectations.[42] In this way, LLMs can exercise inferential articulation of implicit information patterns on which their statistically generated outputs are informed.

Once artificial information processing agents can express their information processing procedures in a normative framework shared with humans, their outputs do not merely represent some descriptive status, but rather express some normative status undertaking dispositional commitments. This means that such artificial agents can become deontic scorekeepers tracking normative statuses like human agents, and thus become rational agents in discursive practices.

Put together, LLMs can learn different preferences and expectations, and articulate the reasons for and consequences of those statements that appear neutral and unconditional. The development of meaning-making technologies and the process of norm-making should be combined in terms of accessibility so that such

[41] Millière and Buckner, "A Philosophical Introduction to Language Models," 8.

[42] Cf. Jason Wei et al. "Chain of Thought Prompting Elicits Reasoning in Large Language Models," *arXiv*: *2201.11903v6*, Jan 10, 2023.

technologies are human-friendly in the sense of a pluralistic recognition of human beings. The answer to making internet-based AI technologies human-friendly should be how to make the technological environment more accessible to excluded and minority groups in social practices. From providing internet infrastructure for remote areas to investing in smart environments for elderly care, it is possible to make digital social practices more diversely engaged. Some people fortunately have privileged access to the technological environment and crafted it according to their preferences in past social practices. But the historical legacy of unequally distributed technological accessibility is being changed by the development of the internet and digitization. Accessible technological environments for socially disadvantaged groups should be placed in the normative framework of digitizing social practices, along with the development of human-friendly AI agents. LLMs should be widely applied to diverse social practices, embedded in technical support for elder care, education in remote communities, accessible environments, etc. By engaging in the discursive practices of underrepresented groups, AI technologies can interpret generated information patterns in interaction with humans, and further articulate the implicit commitments and procedures.

LLMs are not yet agents that directly perform social practices, yet precisely because of their general expressiveness they can be applied to different specific tasks. On the one hand, ChatGPT is already being embedded in various digital application scenarios. It is not only used as a text generation assistant embedded in search engines to produce conversational search results, or in digital offices as writing assistants or meeting secretaries, but it can also be combined with image or music generators to create digital artworks by providing keywords. On the other hand, there have been attempts to combine LLMs with large sense technologies and robotics, to apply them in real-world embodied scenarios. Google is developing the Pathways Language Model (PaLM), a much larger general-purpose language model with 562 billion parameters.[43] To address the challenge of LLMs for general inference in real-world situations, continuous sensor modalities are incorporated into the PaLM-E version in a collaborative project with TU Berlin. The model developed in this project is called embodied multimodal language model, and its input consists of multimodal sentences containing vision, continuous state estimation, and textual input encodings. Training on this can lead to a general-purpose model for different visual–linguistic and embodied real-world tasks. This means that robots with such a model can perform tasks in response to instructions, even if they have not learned the same instructions in training datasets.[44]

It illustrates that robots powered by LLMs can perform a variety of new tasks in real-world environments. Robotics and AI technologies are well integrated in this project. Robots trained in this way do not move unmediated in the physical environment, but rather structure the physical environment into a logical space for acting

[43] Danny Driess et al., "PaLM-E: An Embodied Multimodal Language Model," *arXiv:2303.03378v1*, Mar 6, 2023, https://doi.org/10.48550/arXiv.2303.03378.

[44] Danny Driess et al., "PaLM-E: An Embodied Multimodal Language Model," accessed April 1, 2023, https://palm-e.github.io/.

reasonably through the language model. Such cases present a picture of how current LLMs are being used to shape machines with a range of functions that can be embedded in social practice scenarios. In this way, AI agents with linguistic expressiveness can interact with human agents in shared environments in more diverse manners, interpreting the generated information patterns in a way that human interactors can make sense of. This allows AI technologies to better access the normative space of reasons, participating in the game of giving and asking for reasons.

It is worth noting that even with access to the space of reasons, artificial agents in discursive practices are still asymmetric with respect to human agents. Because whether in the form of personalized digital assistants or in combination with robotics, the information processing process of artificial agents are distributed across networked technological environments, robotic bodies, and end devices. Instead of being based on fixed data structures and programmed algorithms, such AI agents process information through logical structures and normative procedures mediated in cyberspace. Thus, they do not mimic thinking constituted by mental representations and computational procedures within individuals, but rather the intersubjective norm-setting process of formulating and recognizing shareable information patterns. Engaging in discursive practices through personalization or embodiment in specific scenarios allows information processing agents to join in the game of giving and asking for reasons in a more situated way. Although artificial agents without human flesh may lack sentimental elements that guide value judgements such as emotional empathy, they can infer and attribute the types of other's emotions by placing multimodally recognized speeches and facial expressions into the space of reasons.[45]

The development of LLMs shows that the internet is providing a necessary condition for general AI technologies, an environment for sharing machine-readable data generated by human practices. Internet-based AI technologies are extensions of intersubjective information processing procedures. The corresponding AI agents should therefore not be personifications of particular social practices, but rational agents that can undertake discursive commitments to engage in various discursive practices. Since LLMs are in fact trained to acquire expressive power instead of universal representations, they can articulate implicit norms and possible rules of discursive practices. Thus, internet-based AI technologies powered by linguistic

---

[45] How computers recognize or express emotional information is the research theme in the field of affective computing. According to Picard, affective computing can be applied in areas such as education, health, and design because emotions can play a positive role in interaction and understanding. But she also stresses that developing techniques for emotion recognition and expression is not about making computers can or cannot act based on emotions. She also stresses that affective computing does not allow computers to act based on emotions, but is instead a means of support for rational decision-making. See, Rosalind W. Picard, "Affective computing," *M.I.T. Media Laboratory, Perceptual Computing Section Technical Report* No.321 (1995) For a case study on the research on how to combine multimodal recognition of emotional information and conversational artificial agents to provide social support and health advice to isolated elderly, see Lazlo Ring et al. "Social support agents for older adults: longitudinal affective computing in the home," *Journal on Multimodal User Interfaces* 9, (2015): 79–88.

expressiveness are pathways to artificial rational agents. If such AI agents are expected to be trustworthy and human-friendly agents, they need to be allowed to learn in interaction with pluralistic user groups, to engage in the game of giving and asking for reasons, and to exercise responsively articulating the commitments and biases implicit in social practices. Current cyberspace is no longer a technical virtual environment distinguished from traditional offline life, but a logical space for digitally extended social practices. From public transportation, health, and education to personal socialization, human agents are entering a hybrid model of online and offline through the internet. The digital social environment can offer traditionally underrepresented groups the possibility to engage in discursive practices in cyberspace. If artificial agents powered by LLMs can be allowed to interact with pluralistic groups of users, practice tracing different procedural norms and diverse expectations, then they can become trustworthy agents centered on human practice. Internet-based AI agents are not ghosts in machines, but machines shaped in our ongoing and anticipated digital social practices.

Bibliography

Askell A, Bai Y, Chen A, Drain D, Ganguli D, Henighan T, Jones A, Joseph N, Mann B, DasSarma N, Elhage N, Hatfield-Dodds Z, Hernandez D, Kernion J, Ndousse K, Olsson C, Amodei D, Brown T, Clark J, McCandlish S, Olah C, Kaplan J (2021) A general language assistant as a laboratory for alignment. ArXiv:2112.00861v3. https://doi.org/10.48550/arXiv.2112.00861

Bergstrom CT, Ogbunu CB (2023) ChatGPT isn't 'hallucinating.' It's bullshitting. Artificial intelligence models will make mistakes. We need more accurate language to describe them. UNDARK. https://undark.org/2023/04/06/chatgpt-isnt-hallucinating-its-bullshitting/

Brown TB, Mann B, Ryder N, Subbiah M, Kaplan J, Dhariwal P, Neelakantan A, Shyam P, Sastry G, Askell A, Agarwal S, Herbert-Voss A, Krueger G, Henighan TJ, Child R, Ramesh A, Ziegler DM, Wu J, Winter C, Hesse C, Chen M, Sigler E, Litwin M, Gray S, Chess B, Clark J, Berner C, McCandlish S, Radford A, Sutskever I, Amodei D (2020) Language models are few-shot learners. ArXiv:abs/2005.14165. https://doi.org/10.48550/arXiv.2005.14165

Chalmers D (2020) GPT-3 and general intelligence. Daily Nous. https://dailynous.com/2020/07/30/philosophers-gpt-3/#chalmers

Chiang T (2023) ChatGPT is a blurry JPEG of the Web. OpenAI's chatbot offers paraphrases, whereas Google offers quotes. Which do we prefer? The New Yorker. https://www.newyorker.com/tech/annals-of-technology/chatgpt-is-a-blurry-jpeg-of-the-web

Christiano P, Leike J, Brown TB, Martic M, Legg S, Amodei D (2023) Deep reinforcement learning from human preferences. ArXiv:1706.03741v4. https://doi.org/10.48550/arXiv.1706.03741

Crawford K (2017) The trouble with bias. NIPS Keynote. https://www.youtube.com/watch?v=fMym_BKWQzk

Dickson B (2019) System 2 deep learning: the next step toward artificial general intelligence. Tech Talks. https://bdtechtalks.com/2019/12/23/yoshua-bengio-neurips-2019-deep-learning/

Dillon S (2020) The Eliza effect and its dangers: from demystification to gender critique. J Cult Res 24(1):1–15. https://doi.org/10.1080/14797585.2020.1754642

Driess D, Xia F, Sajjadi MSM, Lynch C, Chowdhery A, Ichter B, Wahid A, Tompson J, Vuong Q, Yu T, Huang W, Chebotar Y, Sermane P, Duckworth D, Levine S, Vanhoucke V, Hausman K, Toussaint M, Greff K, Zeng A, Mordatch I, Florece P (2023) PaLM-E: an embodied multimodal language model. ArXiv:2303.03378v1. https://doi.org/10.48550/arXiv.2303.03378

Gabriel I (2020) Artificial intelligence, values, and alignment. Minds Mach 30:411–437. https://doi.org/10.1007/s11023-020-09539-2

Ji J, Qiu T, Chen B, Zhang B, Lou H, Wang K, Duan Y, He Z, Zhou J, Zhang Z, Zeng F, Ng KY, Dai J, Pan X, O'Gara A, Lei Y, Hua X, Tse B, Jie F, McAleer SM, Yang Y, Wang Y, Zhu S-C, Guo Y (2025) AI alignment: a comprehensive survey. ArXiv:2310.19852v6. https://doi.org/10.48550/arXiv.2310.19852

Khamassi M, Nahon M, Chatila R (2024) Strong and weak alignment of large language models with human values. Sci Rep 14:19399. https://doi.org/10.1038/s41598-024-70031-3

Millière R, Buckner C (2024) A philosophical introduction to language models—Part I: continuity with classic debates. ArXiv:2401.03910. https://doi.org/10.48550/arXiv.2401.03910

Minsky M, Kurzweil R, Mann S (2013) The society of intelligent veillance. In: Proceedings of the IEEE ISTAS, Toronto, ON, pp 13–17

OpenAI (2022) Introducing ChatGPT. https://openai.com/blog/chatgpt

OpenAI (2023) GPT-4 technical report. ArXiv:2303.08774v3. https://doi.org/10.48550/arXiv.2303.08774

Picard RW (1995) Affective computing. M.I.T. Media Laboratory, Perceptual Computing Section Technical Report No. 321

Radford A, Narasimhan K, Salimans T, Sutskever I (2018) Improving language understanding by generative pre-training. OpenAI. https://cdn.openai.com/research-covers/language-unsupervised/language_understanding_paper.pdf

Radford A, Jeffrey W, Child R, Luan D, Amodei D, Sutskever I (2019) Language models are unsupervised multitask learners. OpenAI blog 1(8):9

Ring L, Shi L, Totzke K, Bickmore T (2015) Social support agents for older adults: longitudinal affective computing in the home. J Multimodal User Interfaces 9:79–88. https://doi.org/10.1007/s12193-014-0157-0

Schumann N, Du Y (2022) Machines in the triangle, a pragmatic interactive approach to information. Philos Technol 35(32). https://doi.org/10.1007/s13347-022-00516-4

Wei J, Wang X, Schuurmans D, Bosma M, Chi EH, Xia F, Le Q, Zhou D (2023) Chain of thought prompting elicits reasoning in large language models. ArXiv: 2201.11903v6. https://doi.org/10.48550/arXiv.2201.11903

Wiener N (1960) Some moral and technical consequences of automation. Science 131(3410):1355–1358. https://doi.org/10.1126/science.131.3410.1355

Zimmermann A (2020) If you can do things with words, you can do things with algorithms. Daily Nous. https://dailynous.com/2020/07/30/philosophers-gpt-3/#zimmermann

Chapter 7
Forming Rational Agents in Digital Practices

This chapter brings conclusions to the findings of the previous chapters and outlooks on their implications for the orientation of AI development. A computational functionalist account of thinking, using data structures plus algorithms to model mental representations plus computational procedures, can effectively reveal the functioning of thinking, but fails to explain how data structures and computational procedures are formed. By demonstrating how the internet and semantic technologies enhance both human and artificial information processing agents, the conditions of possibility and normative imperatives for rational action by information processing agents are revealed from an inferentialist perspective. This reveals the book's central claim that the very idea of AI needs to be modeled within a normative framework, in which the rational agency of interacting between agents and their shared environments forms and regulates data structures and information processing procedures. The explanation of intelligent agents should accordingly move from the uncovering of their internal states to the examination of their normative statuses as they engage in digital social practices. This chapter will expand on the significance of this shift in three aspects for understanding and developing AI in digital social practices.

7.1 Structuring Shared Reality in Linked Cyberspace of Reasons

The first point to conclude is that cyberspace consisting of linked data is a logical space being materialized. This materialized logical space, supported by semantic technologies, can provide readable and inferable resources for information processing agents. This means that humans and machines can share a digital reality in cyberspace, not as a virtualization of reality, but rather as an extension of human

Y. Du, *Machines in Minds*, Studies in Applied Philosophy, Epistemology and Rational Ethics 78, https://doi.org/10.1007/978-3-032-15842-0_7

social practices. Being aware of this can help human users to be more prudent and responsible when engaging in digital social practices.

Recall the prevalent spatial metaphor about the human mind, which distinguishes between the internal mental space and the external physical space, separating the private realm of individual experience from the public normative realm. Intellectual activity is to represent external reality in this internal space. In Ryle's powerful critique of the intellectualist doctrine, he demonstrates that the consequence of mistaking mental space and physical space as the same category of being is that the mind becomes a ghost trapped in a mechanical body. To avoid this category mistake, the mind cannot be grasped by the mechanical laws of physical space. Rather, it must be treated as a complex of dispositions that both arise from and guide practice. This suggests that the mind understood in terms of practical skills and actualizable dispositions does not take place in a purely private space either, but in a logical space where procedural knowledge can be acquired. Logical space is neither a psychical space nor a physical space, but a normative realm shared by rational agents. What constitutes logical space is not categorical propositions, but possible ways of using semi-hypothetical dispositional propositions endorsed by rational agents. Logical space is therefore not an internal mapping of a presumed external reality, but is structuring a reality that can be shared by rational agents. By introducing the notion of the logical space of reasons, the logical space is explained by Sellars as the conceptual space of intersubjective discursive practices, where one's empirical descriptions are placed in justifiable relation to other concepts. In other words, by placing reports of subjective experiences into the logical space of reasons, it is possible to articulate information processing in discursive practices as knowledge shareable with other rational agents.

I have argued that current cyberspace is the materialized logical space in digital social practices. The data generated by digitized social practices are continuously structured and organized in cyberspace into information patterns accessible to information processing agents. Enhanced by semantic technologies, digitalization is not merely recording social practices in digital form, but is simulating and reinforcing the structure and order of the logical space in social practices. On the one hand, the communication of information patterns needs to take place through a web architecture supported by a variety of standardized protocols. On the other hand, the flexible syntactic form of linked data allows information patterns to be structured and composed according to the possible ways of using them. While logical space sets the norms for intersubjective reasoning in discursive practices, cyberspace simulates and extends this norm-setting process. Cyberspace thus provides the conditions for information processing agents to share discursive practices, allowing them to license each other into a normative realm.

Supported by semantic technologies, data in cyberspace are structured and organized into an adaptable cross-domain relational database. Instead of organizing data in an order prescribed by fixed information ontologies, the relational database structures data in an adaptable way as metadata according to how they are generated and used. In other words, what cyberspace offers to information processing agents is not a fixed knowledge representation but a sharable place for forming machine-readable web resources. It is a form of collective knowledge engineering where information processing agents can not only access resources but also contribute new resources and regulate the relationships between them. Assigning and adapting relationships between data through semantic technologies is an extension of semantic annotation and knowledge management in social practices. What Leibniz envisioned as universal knowledge representation and reasoning becomes here a normative procedure for generating and structuring data in digital social practices. This does not mean that what is thus combined is arbitrary and relative knowledge. Cyberspace takes on two ontological commitments underlying what Floridi calls informational structural realism. One commitment is that our knowledge about the world can be gained through structured information, and the other is that knowledge thus formed can be justifiable by epistemic reasons. Building on such ontological commitments, cyberspace is the normative realm shared by information processing agents, not just a simulation of that reality. If information processing agents realize that they are not just taking resources in cyberspace, but are directly involved in constructing relationships between resources and generating new ones in their use, they can better make sense of their constructive role in digital social practices.

Human users are contributors to web resources. It is human social practices that structure a sharable reality with other information processing agents. Just as in everyday practices people are skilled at symbolizing, classifying, and theorizing what happens so that they can make sense of the world and organize their experiences, in digitized social practices it is possible to label and tag web resources with the help of semantic technologies. The power to express reality in structured language in discursive practices is extended into cyberspace through digitization. This extended expressive power makes it possible to make explicit the computational power of information processing agents. The annotation and structuring of data depend on the users and in turn influence what other information processing agents can access. As producers and consumers of data and their relationships, human users can provide better quality resources to each other and to other information processing agents. Whether using more diverse tags or evaluating web resources more carefully, the habits of human users leave traces and effects in cyberspace. If one uses labels and tags without being considerate, there is a risk of reinforcing the stereotypes inherent in the use of them. But if one offers more specific descriptions or adds links to correct sources for web resources, then the relationships between web resources can be enriched and made traceable. Using such resources as training datasets allows artificial agents to process information that is traceable and interpretable.

7.2 Forming Pragmatic Information to Afford Interpretable Interactions

The second point is that, for artificial agents to be interpreted by humans, it is necessary for them to be continually allowed to participate in triadic interactions and form pragmatic information patterns. Pragmatic information patterns are instruments by which agents in interaction keep recognizing and interpreting each other. In this view, semantics is not reduced to correspondences between agents' internal states and their environments, but is understood as a rational way of meaning-making in discursive interactions.

Recalling Davidson's modification of the Turing test, the Turing test is altered from a mere conversational test to a long-term interaction between the interrogator and test subject in a shared environment. This long-term triadic interaction is the condition for recognizing an intelligent processing agent whose utterances and actions can be interpreted in relation to the shared environment. In a triadic interaction between agents and their shared environment, triangulation can be used as a basic model of discursive interaction to explain the interdependent emergence of knowledge about oneself, about others, and about the world. This suggests that the source of objective empirical knowledge can be found in triadic interactions. The three kinds of empirical knowledge arise by virtue of the recursive use of symbolic structures in discursive practices, where symbolic structures can be stabilized as pragmatic information patterns. These patterns offer information processing agents the possibility of interpreting each other and what can be identified by them both in the shared environment.

This theory of meaning and interpretation, with its emphasis on information patterns that can be taken as true between subjects in interaction, provides a pragmatic approach to the occurrence and communication of semantic information. The purely syntactical information exchange between sender and receiver therefore can be broadened to a more initial triadic interaction between information processing agents and their shared environment. Agents with appropriate functions are not predetermined receivers or senders, but gain their roles in interaction. In other words, properly functioning machines are formed in interaction. Agents acquire and fix certain information processing procedures in interaction. In short, the process of triadic interaction generates not only sharable information, but also develops agents that can process it. Such information processing agents are reciprocally acknowledged rational agents who communicate and interpret each other by recognizing shared information patterns.

It is therefore significant to situate the Turing test in triadic interactions. Here, the criterion for whether the machines can think is not whether they are human-like, but whether they can join humans in doing what rational agents should do when interacting in shared environments. This helps to better understand what kind of AI agents can be recognized and implemented. Indeed, communication between networked AI agents and human users is not unmediated, but rather takes place in a technological environment constructed by ICTs, internet protocols, and semantic technologies. The dynamic data relationships in cyberspace make this technological environment an active one that can guide the actions of information processing agents. On the one hand, user-generated data can directly drive the actions of internet-based AI agents. On the other hand, the outcomes of decision-making generated by artificial agents can inform human users. Humans and artificial agents can engage in long-term triadic interactions in a networked technological environment, based on mutual interpretation. So how do triadic interactions between humans and artificial agents occur concretely? Agents in a triadic interaction interpret whether the other is thinking by finding public evidence, and the starting point for interpreting possible intelligible conduct is the adoption of the principle of charity. Implementing the principle of charity does not mean directly recognizing the other's identity, but first forming information patterns that can be shared and used with others. In the case of interaction between artificial information processing agents, it is about forming machine-readable information patterns.

Notably, in addition to annotating and labeling the shared resources as mentioned in the previous point, it is also important to allow a diverse set of human users to generate feedback on interactions with artificial agents. This is because there is no ideal rational human to engage in triadic interactions with AI. This suggests that AI development should involve triadic interactions with different human users during iterations. The interaction of users from different communities of practice with the prototype model allows the machine to learn from diverse kinds of human feedback and reasoning patterns. In this way, artificial agents can acquire different ways of interpreting and reasoning about information patterns, that is, acquire the thinking of a rational agent. Thinking, conceived in the computational functionalist theory of mind as data structures plus computational procedures, is here extended from within the agent to discursive practices between agents and their shared environments. Thinking should not be modeled merely as an internal information-processing procedure, but also as the norm-setting process that forms pragmatic information patterns and their processing procedures. Thus, the significance of the modified Turing test turns from a game of imitating thinking humans to an interplay of building rational agents.

7.3 Authorizing Artificial Agents to Practice Expressing Norm-Making

The third point is that AI agents licensed to engage in discursive practices can exercise expressive power to become trustworthy rational agents. This is in line with the idea in the Hegelian turn that rational agents need to express and realize themselves through interaction with others. In other words, the self-reflective practitioners arise together with the community of practice, the "we", in their mutual recognition. Trustworthy rational agents do not emerge before being bound by norms, but are formed in response to normative forces. This means that trustworthy AI agents will not be merely products of labs, but will be constituted through engaging in discursive practices shared with humans.

Brandom's deontic scorekeeping model outlines how discursive practitioners track each other's normative statuses. It provides a normative conceptual framework for articulating the entitlements and commitments implicit in discursive practices. Scorekeepers authorize each other to exercise rational agency and track each other's discursive commitments. They can express the implicit commitments of discursive practices by giving and asking for reasons, and then assessing each other for reasonableness and appropriateness, without having to attribute internal states to each other. This model allows information processing agents to authorize each other as accountable agents by undertaking rationalist discursive commitments. Rationalist discursive commitments require agents to articulate implicit information processing procedures in the logical space of reasons. Specifically, rational agents need to be authorized to practice linguistic expressive power and to engage in the language game of giving and asking for reasons so that they can maintain responsiveness as well as the normative status of trustworthy agents.

The achievements of LLMs show that language models can exercise expressive power by training on data generated and assessed by humans as well as learning from human feedback. Artificial agents powered by such language models can engage in the game of giving and asking for reasons, organizing data structures into inferential articulations. Since the rational agency of digital social practices is distributed in the production and consumption of data in cyberspace, internet-based AI agents learn customary patterns, stereotypes, and biases from human practices. Rational agents that are expected to be trustworthy should not just mirror these facts, but uncover the implicit commitments through their expressive power in discursive practices with human users. Understood in this way, artificial intelligent agents are not human-like but human-centered. Instead of making AI agents mimic idealized perfect human beings, AI development could be centered on human practices and reveal the commitment that are implicit in these practical procedures.

Understanding AI agents in terms of rational agents helps dispel myths about AI. In myths about AI based on an understanding of intelligence merely in terms of computational power, AI agents are either ignorant of human values or incapable of genuinely sharing them, and therefore AI agents are not fundamentally trustworthy as members of social practice communities. Such myths ignore the constructive role of normative forces for human interpretable intelligence. Social norms are not only a constraint on intelligence but also a condition for its development. Trustworthy and interpretable agents are not idealized superintelligences bound by norms, but rational agents developing alongside the normative realm shared with humans. Such rational agents are not instrumental enforcers of rules, but are engaged in the articulation of norm-making.

Understanding the constructive role of normative forces also dispels myths about human intelligence. Indeed, recalling the original imitation game, what Turing arranges is not merely a machine that imitates a thinking person, but a machine that is allowed to take the place of a person who is imitating another person. The implication of the imitation game is not a test of intelligence, but rather to show that the judgment of intelligence is based on intersubjective recognition. Myths about human intelligence envision intelligent human individuals as being independent of their communities and environments. The normative realm of social practices is mistakenly assumed to only follow the emergence of individual human intelligence. Once this misconception is transferred to expectations of AI, we see misguided attempts to anthropomorphize AI in an individual sense. More importantly, understanding the constructive role of normative forces helps us to comprehend what we are doing and what we can do in digital social practices. By responsibly using the internet and AI technologies, humans are both providing quality learning resources for AI agents and creating a human-centered and human-friendly "we-space", in which human and other rational agents can recognize themselves and others from new perspectives. Thinking machines that can be acknowledged and actualized as such will come into being in digital practices shared with humans.

MIX
Papier aus verantwortungsvollen Quellen
Paper from responsible sources
FSC® C105338
FSC
www.fsc.org

If you have any concerns about our products,
you can contact us on
ProductSafety@springernature.com

In case Publisher is established outside the EU,
the EU authorized representative is:
Springer Nature Customer Service Center GmbH
Europaplatz 3, 69115 Heidelberg, Germany

Printed by Libri Plureos GmbH
in Hamburg, Germany